Praise for

Earthen Floors

At last! The mystery is removed from most wonderful floors in the world. First thing you do: go find an earthen floor, pull off your shoes and socks, have a little walkabout, and discover great surprise and tactile delight. Next thing you do: buy this book, learn everything you need, and go make one yourself. Thank you Sukita and James for making this ancient treasure accessible to all.

— Bruce King, PE, Director, Ecological Building Network

Earthen floors have long been shrouded in mystery... Natural building enthusiasts have known about them and lusted after them, but despite the vast amount of natural building information produced in the past two decades, a tested and true treatise on earthen floors has been missing. No longer! This book is full of information, hard-won lessons learned and the real heart and spirit of natural building. The authors successfully de-mystify the process of mixing, installing, finishing and maintaining an earthen floor in a way that will help novices and seasoned builders alike to produce floors that are the most beautiful, sensual and eco-friendly option available.

— Chris Magwood, author, *Making Better Buildings* and *More Straw Bale Building*

With *Earthen Floors: A Modern Approach to an Ancient Practice*, Sukita and James have come up with the ideal mix of inspiration, experience, technique and instruction. This book will be of use and benefit to both experienced practitioners and those just dipping their toes into the subject of earthen floors for the first time. It is always a delight to find a well-written, clear and practical guide to natural building practice, and a special treat when fun and informative personal stories are interwoven with good graphics to bring the whole process to life. This book makes a great contribution to the field on every level.

— David Eisenberg, Director, Development Center for Appropriate Technology

This is an excellent resource. Packed with expert first hand experience and knowledge, it is set to be the bible on earthen floors and fills a much needed gap in this wonderful area of natural building.

— Adam Weismann and Katy Bryce, authors,
Building with Cob and *Using Natural FInishes*

Earthen Floors fills a long-standing gap in the catalog of natural building publications in providing reliable and thorough instruction in earthen floor construction. With a style that is both comfortable and authoritative, this book offers the detailed instruction necessary for executing a quality earthen floor in a variety of different contexts. Crimmel and Thomson have drawn upon their extensive and direct experience in their craft and share valuable lessons learned with their readers. From the exploration of the historical roots of earthen floors to a look at the future of how we can improve upon this practice, *Earthen Floors* is the most comprehensive and thorough resource on earthen floor technology available today. Earthen Floors offers a gift to the natural building community that makes this technology more accessible to DIYers and professionals alike.

— Jacob Deva Racusin, author, *The Natural Building Companion*,
Co-Owner, New Frameworks Natural Building, LLC

This book is a welcome treasure, a long awaited and generous sharing of the detailed knowledge gained by the authors through years of testing and teaching and designing and installing earthen floors. In a thoughtfully presented text, we are now gifted with a compendium of information in one place, rather than the myriad conversations, tips, trials and experiments common in the last years of rediscovering and refining this valuable and timely technology. My clients and I will be indebted to Sukita and James for years to come.

— Laura Bartels, Green Weaver Inc.

EARTHEN FLOORS
A MODERN APPROACH TO AN ANCIENT PRACTICE

EARTHEN FLOORS
A MODERN APPROACH TO AN ANCIENT PRACTICE

SUKITA REAY CRIMMEL AND JAMES THOMSON

new society
PUBLISHERS

Copyright © 2014 by Sukita Reay Crimmel and James Thomson.

All rights reserved.

Cover design by Diane McIntosh.
Cover images: Top photo © Miri Stebivka, middle and lower photos © James Thomson
Back cover: Top photo © Mike O'Brien, Left: © Miri Stebivka, Right: © James Thomson

All interior illustrations by John Hutton.
Interior photos by Miri Stebivka, Mike O'Brien,
James Thomson and Sukita Reay Crimmel, and others as listed.

Printed in Canada. Fourth printing January 2022.

New Society Publishers acknowledges the financial support of the Government of
Canada through the Canada Book Fund (CBF) for our publishing activities.

Paperback ISBN: 978-0-86571-763-3
eISBN: 978-1-55092-561-6

Inquiries regarding requests to reprint all or part of *Earthen Floors* should be addressed to
New Society Publishers at the address below.

To order directly from the publishers, please call toll-free (North America)
1-800-567-6772, or order online at www.newsociety.com

Any other inquiries can be directed by mail to:

New Society Publishers
P.O. Box 189, Gabriola Island, BC V0R 1X0, Canada
(250) 247-9737

New Society Publishers' mission is to publish books that contribute in fundamental ways
to building an ecologically sustainable and just society, and to do so with the least possible impact on the environment, in a manner that models this vision. We are committed
to doing this not just through education, but through action. The interior pages of our
bound books are printed on Forest Stewardship Council®-registered acid-free paper that is
100% post-consumer recycled (100% old growth forest-free), processed chlorine-free, and
printed with vegetable-based, low-VOC inks, with covers produced using FSC®-registered
stock. New Society also works to reduce its carbon footprint, and purchases carbon offsets
based on an annual audit to ensure a carbon neutral footprint. For further information, or to
browse our full list of books and purchase securely, visit our website at: www.newsociety.com

Library and Archives Canada Cataloguing in Publication

Crimmel, Sukita Reay, author
    Earthen floors : a modern approach to an ancient practice / Sukita Reay

Crimmel and James Thomson.

Includes bibliographical references and index.
Issued in print and electronic formats.
ISBN 978-0-86571-763-3 (pbk.).--ISBN 978-1-55092-561-6 (ebook)

    1. Flooring, Earth--Amateurs' manuals. 2. Flooring, Earth--History.
I. Thomson, James, 1977-, author II. Title.

TH2529.E27C75 2014   698'.9   C2013-908693-5

                                      C2013-908694-3

Dedicated to clay, sand and straw,
and their myriad combinations and creations

Contents

Foreword

by Paula Baker-Laporte and Robert Laporte

THE FIRST TIME I LAID EYES ON AN EARTHEN FLOOR I was amazed. I had newly arrived in New Mexico, a young architect from Toronto, and was working on my first adobe renovation. The floor was beautiful! It wore a graceful patina, a subtle trace of more than a hundred years of footsteps. Firm but forgiving, warm, inviting, like walking on leather. It changed me. Flash forward a couple of decades...and now Robert and I have enjoyed earthen floors in our own homes for the last 15 years. In our current home construction, Sukita gave an earthen floor building workshop, and we used her Claylin brand of materials and finishes that she has developed. Living with these floors, it has become obvious to us that she has not only mastered an ancient craft but advanced it, both aesthetically and in durability, from floor of necessity to a floor of first choice. It is always a thrill to introduce someone new to the sensual pleasures of standing barefoot on a finished earthen floor because, more than with any other aspect of our natural home, people "get it" viscerally straight from their soles to their souls.

Paula Baker-Laporte and Robert Laporte.

As with so many ancient building techniques, all but forgotten in a world that has embraced high-tech building, an earthen floor embodies the knowledge of how to build durably, with the materials at hand with the very least impact on our fragile ecosystem. And as with so many natural ways to build, this craft, too, almost became extinct in North America but for the efforts of a few pioneering champions who were willing to explore, experiment and coax information from the few remaining elders who remembered the craft.

Sukita and James are fine examples of this championship. It takes noble character to be willing to learn a craft through countless backbreaking episodes of trial and error. It takes incredible generosity to share the results of this labor of love in a carefully written book so that anyone with the will and the ability to follow instructions can experience success from the get-go.

The book not only describes step-by-step how to craft an earthen floor, it is packed full of advice that could only have been gleaned from a decade of hard work and building enough earthen floors to fill a football field! There is an attention to sharing details like "It can be difficult to remove oily tape once it has hardened" that will save others countless hours of needless labor. It will also bring the "been there, done that" smile to anyone who has ever accomplished an innovation of significance, as the image of Sukita or James painstakingly stripping dried oily tape from walls...only once... comes to mind! As with so many natural building pioneers who have passionately pursued the revival of building ecologically with these time-tested materials, they have had to learn without the benefits of apprenticeship under a learned master, but learn they did.

With this book, Sukita and James can enjoy a well-earned seat in the Natural Building Renaissance Hall of Fame...a future historic structure that has yet to be built of clay, sand, straw, wood and stone. We feel certain that *Earthen Floors: A Modern Approach to an Ancient Practice* will quickly become the bible for earthen floor construction and the solid base upon which the next generation of earthen floor crafters will stand.

Acknowledgments

THIS BOOK IS THE PRODUCT not only of many months of work but also of years of learning and working with earthen floors. We are incredibly grateful to those who have been supportive during the writing process and also over the years of work, providing conversation and play that have given us the knowledge to set out on this endeavor. We are indebted to our teachers, our colleagues, our students, our friends and our clients who have encouraged us, humored us and been patient with us along the way. After more than ten years of this work, their names are too many to list here. We are are grateful to all of you.

In particular, we wish to thank our early teachers, who first encouraged us to get muddy: Coenraad Rogmans, Rob Bolman, Patrick Henneberry and Ianto Evans, thank you for your early inspiration and guidance as we set out on this journey. We are eternally grateful for our many colleagues who have gotten their hands dirty alongside us as we learned and shared insights, and have become dear friends along the way: Lydia Doleman, Bernhard Masterson, Chad Tate, Tracy Thierot, Mark Lakeman, Jeremy Rosenbloom, Tim Kennedy, Chris Magwood, Carey Lien, Kindra Welch, Bobby Wilcox, Alexis Jaquin, Eva Miller, Tim Russell,

Anna Wolfson, Erika Ann Bush, Katherine Ball, Brandon Smyton. An extra thank-you to our friends who have taken time to field our myriad questions, lent us their expertise and answered our calls for personal stories and experiences to include in this text: Michael Smith, Robert LaPorte, Peggy Frith, Anita Rodriquez, Bill Steen and Autumn Peterson.

We consulted with several of our "author" friends and family members as we under took this task, to make sure we were not crazy (we were), and not making too many mistakes (we didn't, thanks to them). Thanks to Daniel Lerch, Peter P. Thomson and Kelley Rogers for giving us their time and wisdom.

This book can only work because of the many beautiful illustrations and photos included alongside the text. We are deeply grateful to John Hutton for his wonderful drawings, images created just from our verbal descriptions; to Miri Stebivka, who took many of the process and finished product photos (including one on the cover); and to Mike O'Brien for generously allowing us to use several of his excellent photographs.

Finally none of this would be possible without New Society Publishers. Thanks are owed to many: Heather Nicholas, who believed in the project early on and encouraged us to submit our manuscript; Sue Custance, the production manager who showed tremendous patience with our seemingly endless tweaks and reworkings; Diane McIntosh (and Sue) for coming up with the beautiful cover design; Scott Steedman for his careful copy editing; MJ Jessen for design and layout of the book; Greg Green for putting together the color section; and Ingrid Witvoet for steering the whole project to a successful finish. New Society Publishers has proven its commitment to making information about natural building and living available to a wider audience, and we are proud to be affiliated with them.

Introduction

"Wow!" is the most common exclamation heard when people set foot on an earthen floor for the first time. It's usually followed by "It's beautiful!" or, more often, "What is it?" There is something foreign and strange about an earthen floor, yet something familiar and comforting. For people who are used to concrete, tile, hardwood or carpet, taking a step onto an earthen floor is a refreshing change.

Most people are unfamiliar with the idea of an earthen floor, and are understandably skeptical. "Aren't they dusty?" they ask. When people imagine an earthen floor, they probably

A beautiful earthen floor in Crestone, CO.
(CREDIT:
JAMES THOMSON)

1

think of a dirt floor in a small shack in a village in a faraway country, or maybe something from the past. While this type of primitive floor can actually be quite comfortable if cared for properly, the earthen floors described in this book are a modern adaptation of these ancient techniques. Today's earthen floors are attractive, durable and excellent options for modern homes.

The basic ingredients in earthen floors are simple: sand, clay and some sort of fiber (usually chopped straw). Other additives may include pigments for color and manure for extra fiber. Finally, once applied and fully dried, the floor is sealed with coats of drying oil and wax.

People love the unique look and feel of earthen floors. They come in a range of dark, rich earth tones, often flecked with lightly colored fibers. They feel different, too: walking barefoot over an earthen floor is a starkly different experience from walking on a concrete floor. The first obvious difference is temperature: earthen floors feel warmer to the touch. Second, an earthen floor is softer than a concrete floor. The difference is subtle, but important. Anyone who has worked a job that required standing or walking on a concrete floor knows the impact on the body of working on a rock-hard

surface. Many have experienced that standing and working on earthen floors is easier on the body. Finally, the minor undulations and irregularities that are a result of the hand-troweling process more closely mimic textures and patterns that we might expect to find in nature and feel more natural to our feet, like walking on the earth.

This book will describe the history of earthen floors, give best practices for where and when to use them and provide

(Credit: Mike O'Brien)

step-by-step instructions for installing a floor from scratch. It is intended to be a practical guide for DIYers, contractors, architects and anyone with an interest in trying their hand at this amazing practice.

The authors (and others in the natural building field) have developed the installation techniques described here over more than a decade of work. This doesn't mean they're the only or even the best way; they're simply a way that has been shown to work over time and in a variety of circumstances. Building is a creative practice. Builders develop their own techniques and methods that work well for them, and readers will likely modify and adapt the suggestions in this book for their own purposes. Building with earth is different from other kinds of building. Because earth is a locally sourced product, it is inherently unpredictable. What works in one location with one type of soil may not work somewhere else. People who choose to pursue expertise in this field should be prepared for surprises and challenges as they learn and gain experience.

No book can teach everything there is to know about any topic, and this one is no exception. The intention is to provide enough information to improve the odds for success, and to prevent readers from making some of the many mistakes the authors have made. But mistakes will invariably happen anyway; hopefully they will be opportunities for further learning and development rather than causes for despair.

The Authors

Sukita Reay Crimmel

I feel connected to my human history working with the raw ingredients I use to make earthen floors: the lineage of humans using what is locally available to make shelter. Like a cook in a kitchen, I feel a relationship with the properties of clay, sand, straw, oil and wax and find simple inner joy in blending these ingredients to create beauty. The process involved in transforming these materials from piles into floors has taken my breath away again and again. The shimmer of the clay, the strength of the sand, the color of the

Sukita Reay Crimmel.

(CREDIT: MIRI STEBIVKA)

straw and the support of oil and wax delight me. I feel like a connector of our ancient past to our modern present. Something about these floors feels sacred, especially when the materials come from our own digging at the location we are building. The experience of community among those installing the floor often brings out joyful play. The floors take on all this joy of creation and ongoing admiration. What a beautiful foundation to life!

At college I studied architecture. The University of Oregon was known for its green building program, and through my coursework and involvement with a student group, I found out about natural building. The simple beauty of earthen architecture got into my heart and lit my passion! Hands-on experience was to be my teacher for some time, learning how these materials performed beyond the embodied energy numbers and thermal qualities I had learned in school. After a couple of years working for a progressive builder in Eugene and engaging in the natural building community on the West Coast, I moved to Portland, Oregon, in 2002 and began to engage with natural building in an urban environment.

I started a contracting business (From These Hands) that specialized in earthen plasters, earthen floors and other earthen features. In 2007, my focus on earthen floors was turned up a notch with press coverage of my work in the *New York Times*.

In 2011, I started Claylin LLC, a manufacturing company that produces a ready-mix earthen floor blend, as well as sealing oils and a finish wax. I offer these products to ease the questions and challenges of new recipes and to bring earthen building materials into a market that is unfamiliar with them. It has been my mission to rekindle the connection, knowledge and confidence to build with the earth. And it is my intention with this book to help those who will create their own recipe or use Claylin to have the support of my years of learning. It brings me great joy to collaborate with my friend and colleague James Thomson in writing this book.

James Thomson

When I was eighteen, I took my first trip to the Pacific Northwest. Growing up in New

James Thomson.
(Credit: Miri Stebivka)

England, I had spent plenty of time in the woods, but the woods of the Northwest were something different entirely. On the Olympic Peninsula, towering trees and mountains dominated the landscape for as far as the eye could see. But there was something else unfamiliar, too: clear-cuts. These huge scars on the landscape hit me like a punch in the gut. It was easy to make the connection between the conventional construction industry and the havoc it wreaks on our fragile environment.

I first became interested in natural building as a way to help preserve these (and other) forests. I wanted to learn to build in a less destructive way, with materials that were all around us and whose use would leave a smaller scar on the landscape. In 2004, I moved to Southern Oregon to learn to build with earth and straw, and have been "covered in mud" ever since.

I soon realized that natural building is not just about swapping one material (earth) for another (wood/concrete/vinyl). It is about fundamentally re-evaluating how we live, rather than just what we live in. It's about changing our relationship to the structures that shelter us, from one that is mostly financial in nature to one that is more holistic, that values how we really want to live over how much our investment grows.

Building with earth is one way to help us deepen our connection with our homes. The unique aesthetic it offers reminds us of the larger Earth outside our door. It's also a technique that leaves a light footprint on the environment and doesn't bring a lot of toxic junk indoors.

And best of all, earthen building techniques are accessible to novice builders. Like all things, they require practice to develop skill, but the materials are safe and forgiving and, perhaps more importantly, fun to work with. I find that most people have an innate connection with working and playing with earth. Maybe it's from early memories of making sandcastles on the beach or mud pies in the backyard on a hot summer day. Or maybe it's because these materials are all around us; they make up our world, they make up *us*.

I poured my first earthen floor in the summer of 2004. I couldn't believe that such simple materials could make such a sturdy floor; even after years of installing them, I'm still sometimes surprised that they work so well. Earthen floors are a great starter project for budding natural builders. The process is similar to concrete floors, as is the aesthetic, and many find it more accessible and realistic than more radical techniques like cob or straw bale. And best of all they can be installed in most preexisting homes, with minimal alterations to the existing structure. I feel grateful to have connected with Sukita Reay Crimmel several years ago; working with her has greatly increased my understanding of and appreciation for earthen floors. Together we hope to demystify this amazing technique and make it available to you, now. I hope you'll welcome some earth into your home with an earthen floor!

Sukita and James join forces...

Writing this book has been a truly collaborative process. We began in late 2011, but had already established a friendship and working relationship by then. Our writing collaboration started with the instruction manual for Sukita's earthen floor product Claylin. Creating this manual made us realize how much experience and knowledge we had collectively accumulated, and wetted our appetite for writing a more general-audience book.

In the spring of 2011, Sukita met Heather Nicholas from New Society Publishers at the *Mother Earth News* fair in Puyallup, WA. Heather encouraged Sukita to write a book about natural building; the obvious choice was earthen floors. Realizing that this would be a big undertaking, Sukita asked James to collaborate on the project, and he agreed.

We have each brought our own skills and knowledge to this project. Sukita has established herself as a nationally renowned earthen floor expert, and has installed over 20,000 square feet of earthen flooring. James has brought not only his own hands-on experience but also an eye for writing and editing. We've spent many, many hours together over the past two years, and can happily report that we still get along well! It has been a near-perfect collaboration, with each of us feeling like we could not have done the project without the other. Over the months, the information in this book has been written and rewritten, with the hope of creating a clear, straightforward and definitive guide to all things earthen floor.

SECTION 1

ABOUT EARTHEN FLOORS

The idea of living on a floor made of earth is not new; humans have been doing that for centuries. What is new is the idea of bringing these ancient materials into our modern homes. It is valuable to understand where the techniques used today came from, and how earthen floors are different from (and similar to) other flooring options, in order to make sure they are installed in appropriate settings. (Photos: Left: James Thomson, Right: Miri Stebivka)

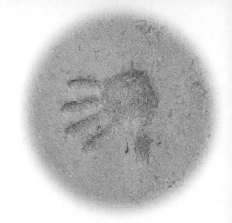

CHAPTER 1

The History of Earthen Floors

B<small>UILDING WITH EARTH HAS BEEN A STAPLE OF HUMAN</small> <small>CIVILIZATION</small> for millennia. Confronted with the question of how to shelter themselves, our ancestors looked around and found an abundance of natural building materials right under their feet.

600-year-old "Hakka" communal earthen structure, China. (C<small>REDIT</small>: © L<small>IUMANGTIGER</small> | D<small>REAMSTIME.COM</small>)

Using soil, sand, rocks and plant fibers, they built durable and comfortable buildings that would stand for generations. They used these same materials to add smooth, attractive finishes, providing bright, beautiful and durable interiors and exteriors. For most of human history, people have lived in houses that they (or a family member) had built from materials found close by. In many parts of the world, these ancient traditions are still in use today.

Earth is abundant and ubiquitous. Trees don't grow everywhere, but the earth is always underfoot. There are several earthen building techniques that readers may be familiar with. The most widely known is sun-baked clay bricks, often referred to as adobe. There is evidence that ancient Egyptians made mud bricks more than four thousand years ago (2000 BCE). Every continent on Earth has ancient structures built from mud or clay bricks. Here in the US, there are many examples of ancient adobe buildings, including the oldest continually occupied structure in North America: Taos Pueblo in New Mexico, about a thousand years old. Other common earthen building techniques are waddle and daub (mud plaster smeared over a matrix of woven sticks), cob (similar to adobe, but without the bricks) and sod (bricks of earth and roots cut from the earth). Earthen plasters, sometimes mixed with manure and natural pigments, have been utilized to beautify and preserve earthen dwellings. Human societies the world over have employed these (and similar) techniques for thousands of years to create shelter. And these techniques are still in use today; it is estimated that a quarter to a third of the world's population lives in houses built partially or entirely of earth.[1]

During the industrial revolution and the population (and building) boom that followed, many of these traditional techniques were displaced by new materials and building designs, such as lightweight wooden structures framed with wooden studs that could be erected quickly and cheaply. A new industry of building product manufacturers arose to satisfy the insatiable demand for more houses. By "outsourcing" much of the materials preparation to these manufacturers, buildings could be built even faster.

This once-new paradigm has become the standard in developed nations. Building products manufacturers, from lumber companies to paint companies, continue to make and sell products that allow buildings to be completed in a fraction of the time it would have taken our ancestors just three or four generations ago.

Yet this time saving has a cost: not only a financial cost but also a cost to the health of our planet and our bodies. The buildings industry is one of the most destructive forces on the planet, producing toxic chemicals that we are encouraged to put in our homes (paint, wood treatments, formaldehyde, glues)

Wooden stud framing in progress. (Credit: © Picstudio | Dreamstime.com)

Scars on the landscape left behind by logging operations. (Credit: © Steve Estvanik | Dreamstime.com)

Buildings Are the largest contributor to climate change

(adapted from Architecture 2030: Climate Change[2])

According to the US Energy Information Administration (EIA), the building sector consumes nearly half (48.7%) of all energy produced in the United States, about the same amount of energy consumed by transportation (28.1%) and industry (23.2%) combined. Globally, these percentages are even greater. Most of this energy is produced from burning fossil fuels, making the building sector the largest emitter of greenhouse gases on the planet — and the single leading contributor to anthropogenic (human-created) climate change. Nearly half (46.7%) of all CO_2 emissions in 2009 came from the building sector (transportation accounted for 33.4% and industry just 19.9%). With so much attention given to transportation emissions, many people are surprised to learn these facts.

and leaving vast swaths of clear-cut or strip-mined land in its wake.

The environmental impact of a building doesn't end when the last contractor leaves the building site, either. As the evidence for climate change becomes overwhelming, the calls for changing the status quo are growing louder.

This book is not intended to address all of these challenges, but it does pose a possible answer to the question "Is there a better way?" The Earth provides an incredible abundance of durable, non-toxic and beautiful building materials that can be found almost anywhere. Our ancestors knew this. Can we learn, again, how to use them?

The rise of "green building"

In the 1970s, rising awareness of environmental pollution coupled with an energy crisis spurred the growth in North America of what has come to be known as the green building movement. Green builders strive to improve the energy efficiency of new buildings and to use materials that are less toxic and more sustainably produced. Adhering to the philosophy that the "greenest" building is the one that is already built, they also work to retrofit existing structures for improved energy efficiency as well as creating spaces that are more pleasant to work and live in.

A whole industry has arisen to address the concerns of those who struggle with the destructive nature of the conventional construction industry. Green building, from whole building design to remodeling materials, is on the rise.

Natural building

"Natural building" is a relatively new term that describes a building philosophy that emphasizes sustainability through using minimally processed, locally available, plentiful and renewable resources to create healthy living environments. Natural builders value handmade and site-processed materials over store-bought products and favor small, thoughtful buildings over large, extravagant ones. Green buildings tend to rely more on green building products that have been manufactured to meet the growing needs of the industry, while natural building eschews products when locally available materials will suffice. Natural building is still a very small fringe movement, but technical books on natural building processes continue to become available, and building codes are being rewritten to be more accepting of natural building practices.

Earth is a favorite building material of natural builders,

A remodeled bedroom in a 1926 Craftsman-style home, with earthen plaster walls and an earthen floor. (Credit: James Thomson)

and earthen building has seen a small renaissance in the last two decades here in the United States and around the world. What started as a movement of homesteaders and off-the-grid DIYers has grown to where it is now possible to find earthen building projects in many urban areas of the most developed countries. Germany, for example, has a small but thriving earth-building culture, where historic and new buildings alike have been updated and built with earthen materials. In Portland, Oregon, a growing natural building movement has brought earth building into the urban environment; there are many public gathering spaces created out of earthen materials, and there are "conventional" houses that have been retrofitted with earthen paints, plasters and of course floors.

Earthen floors are a great option for those who want to bring earth into their homes. The technique is relatively easy to learn, and the floors can be installed in all types of existing buildings. They sell themselves on their aesthetics alone, without even considering their unique feel, low toxicity, minimal environmental impact and thermal benefits. They can be applied in a variety of situations and conditions, and are suitable for most general-use rooms. These are modern earthen floors.

Earthen floors

The concept of an "earthen floor" is not new. Homes have been built directly on the earth for many millennia. As late as 1625, most European houses had a tamped earthen floor, and visitors had to wipe their shoes on an entry mat to make sure they didn't get it muddy or dusty. Early settlers in North America lived directly on the earth, like the many generations of Native Americans who preceded them. As the colonies expanded and industrialized, the abundance of wood available brought about the plank wood flooring of the Colonial Era (1607–1780) and earthen floors fell out of favor.[3] Still, in many parts of the world, it is not hard to find people living on floors made of earth. Sometimes the residents use sealers, to stabilize the earth more permanently; examples of simple, low-cost sealers include ox blood, ghee, vegetable oils and even used motor oil. More typically, traditional earthen floors are just the raw earth beneath the house, tamped down with human feet and moistened frequently with water to keep the dust down.

The modern revival of earthen floors in North America has its roots in the homes and buildings of the native southwest of the United States, which maintained "traditional" (raw) earthen floors into the twentieth century. As people moved into more conventional modern homes, the techniques became less and less used until they were almost lost entirely. Yet a small number of craftspeople continued the practice. The lineage of the "modern earthen floor" can be traced to three early practitioners.

Taos Pueblo, the oldest continuously occupied dwelling in North America, is built from earth.

(CREDIT: JAMES THOMSON)

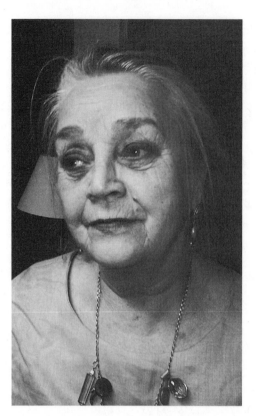

Anita Rodriguez.

(CREDIT: DON ROBERTS)

Anita Rodriguez: Growing up in New Mexico, Anita was surrounded by adobe buildings. Traditionally, the finishes in these buildings, including the plasters and floors, were applied and maintained by women, who were known as *enjarradoras* ("plasterers" in Spanish). In the 1970s, Anita was inspired to learn the trade of earthen plasters and floors, and she sought out the women who still knew the skills. Much of their knowledge was being lost as their communities collapsed and modern construction practices were adopted. Anita took what she learned from these women and continued to experiment for years, integrating earthen materials into her construction business. Anita picked up the use of linseed oil, a widely available sealant, from a colleague and incorporated its use into her practice.

Bill and Athena Steen: The Steens have been integral in spreading the technique. While writing *The Straw Bale House,* they came in contact with Anita. Bill writes:

> [Anita] had come up with a method for doing adobe floors that didn't crack. At that time, most adobe floors were poured three to four inches thick. At that thickness, they typically cracked, largely because adjustments to the site soil were rarely made. Those cracks were often patched before sealing the floor, giving a look like rustic flagstone. The woman with the crack-free

*Bill and
Athena Steen.*
(CREDIT: BILL STEEN)

adobe floors was Anita Rodriguez.... We secured a job for ourselves doing an earthen floor and we wanted to use Anita's method. We compensated Anita for her formula from revenues earned on that job and we were on our way. About the same time we wrote a very simple booklet on the technique, every now and then one finds a copy in some obscure location. The essence of Anita's method were applying two half-inch coats as the finish for the floor and then sealing them with four coats of heated and progressively thinned coats of linseed oil.

Bill Steen:

My favorite floors were always those that were nothing more than plain dirt and were renewed with dampening and sweeping on a regular basis. Those were primarily in Mexico. My mother continued dampening and sweeping the yard around our home as I grew up, a practice common in the patios of many old homes in southern Arizona. It made the earth smell fabulous.

The Steens developed the process further and started using insulation:

> I think one of the major changes we adopted was to switch from using off-the-shelf boiled linseed oil, that is toxic and foul smelling, to using sun-thickened raw linseed oil that we produced ourselves. We also started adding insulation beneath the floors and separating them from the ground with adequate drainage where needed.

And finally, they gave the floors the name we use today:

> We coined the term "earthen floors" in order to make it clearer to a wider audience instead of how they were known in our part of the world, "adobe floors." From that point they seemed to really take off.

Sukita:

I first learned how to make earthen floors from my time working with Robert Bolman in Eugene, Oregon. Robert learned what he knew from Bill Steen. And today as I continue to learn, with lots of trial and error, I am personally inspired and excited to find a connection with Anita Rodriguez and my fellow *enjarradoras*.

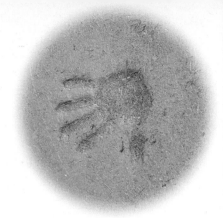

CHAPTER 2

Understanding Earthen Floors

EARTHEN FLOORS ARE VERY VERSATILE and can be installed in a variety of locations, but are not appropriate for every application. It's not possible to write a list of all the places they can (or can't) be used, but understanding their characteristics will help the installer determine what applications are appropriate.

The main characteristics of an earthen floor are:

- Texture
- Flatness
- Hardness
- Weight
- Thermal mass
- "Waterproofness" (or resistance to moisture and stains)
- Toxicity
- Flammability
- Odor
- Embodied energy
- Color
- Thickness

- Installation and drying times
- Cost
- Cracks

Texture (rough, smooth, shiny, matte)

The final texture of an earthen floor will depend on the raw materials used in the mix and how the floor is finished. The raw ingredients of sand, clay and straw can range in size and texture from fine to coarse, depending on where they come from and how they are processed. Coarse, large-grained sand will tend to leave a rougher or more "porous" texture, while finer-grained sand will leave a smoother, "tighter" surface. Similarly, floor mixes with higher clay contents will have a less porous surface (more on clay content in Chapter 5). Most fibers will be visible on the surface; larger and longer fibers will create a more textured surface, while smaller fibers are less noticeable to the eye or touch.

Burnishing, an optional finishing step described at the end of Chapter 8, will create a smoother surface by pressing the sand grains down into the clay matrix and bringing the clay particles up to the top. Skipping this step will leave a more roughly textured floor. Finally, the application of a wax layer can change a floor's finish

A floor with a matte finish.

(Credit: Miri Stebivka)

from matte to slightly shiny. Certain finishing products will result in a glossier finish than others.

Flatness

Earthen floors are hand-troweled, which often leaves subtle highs and lows that are only visible up close. These gentle undulations give earthen floors an organic beauty; they also may mean that the floors are not perfectly flat. This shouldn't be understood to mean that a floor will not be "level."

A potential drawback to the irregular surface is that some four-legged pieces of furniture may wobble a bit (the same effect found at a restaurant with a table that is wobbly because the floor or the table legs are uneven). This is usually easy to solve by shifting the piece of

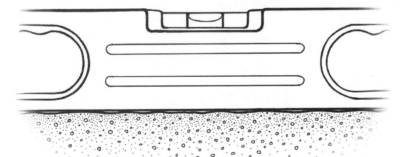

A floor treated with a glossy wax or resin gives a shiny finish. (CREDIT: MIRI STEBIVKA)

A floor may be level, but not perfectly flat! Note the subtle undulations under the level, a result of the hand-troweling process. (CREDIT: JOHN HUTTON)

furniture a few inches in any direction, or by inserting a shim under the wobbly leg.

Hardness

At first glance, earthen floors are sometimes mistaken for concrete floors; they are smooth, continuous and hard. Closer inspection reveals that earthen floors are actually not as flat (see above) or as hard as concrete. Their hardness can be compared to that of a wood floor: harder than fir or pine, but not as hard as oak. The dried finish oil creates a surface that has some plasticity (often likened to traditional linoleum, which was made with linseed oil) and is more likely to dent rather than crack from a point load or the impact of a heavy object.

Due to their relative softness, earthen floors may not be appropriate for a space where one might put (or drop) heavy objects on the floor, such as a tool shop.

James:

I lived in a house that had a section of concrete floor right next to a section of earthen floor; often I would invite guests to remove their shoes and socks and walk on each section. When placed side by side, the earthen floor felt much softer than the concrete floor (not to mention warmer, too). The softness is subtle but enough to make for a more comfortable surface to stand and work on, and is especially noticeable if you stand on it for many hours in a day.

A concrete floor (on the left) meets an earthen floor (on the right). Bare feet can easily tell the difference!
(Credit: Miri Stebivka)

Weight

Earthen floors are made from a mix of sand and clay, usually poured ¾ to 2 inches thick. At 1-inch thick, they weigh about 10 pounds per square foot, comparable to a concrete floor of equal thickness or a thick tile floor. Weight is not important if you are installing the floor on an "on-grade" subfloor (like a concrete pad or compacted gravel; see Chapter 6 for a discussion of subfloors) but is an important factor on a framed wood subfloor.

Most conventional homes today have wooden subfloors, even on the ground floor. This allows for easy insulating under the floor and provides a space for mechanical systems like heating ducts and plumbing. A system of posts, beams and joists with plywood sheets on top creates the subfloor for the finished floor material. This same system is used for other levels in a building, so it is absolutely possible to put an earthen floor on a second story or higher.

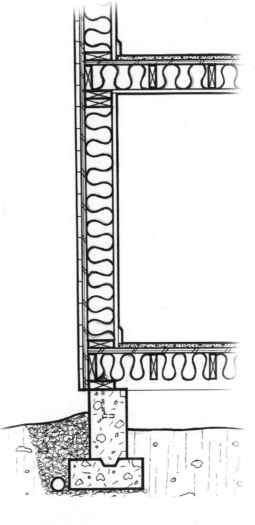

Cross section of a conventionally framed 2-story house, showing wood-framed subfloors on the first and second level.

(Credit: John Hutton)

Putting more weight on a second floor will increase the entire "dead load" on a building's structure and foundation. This term describes the overall weight of the building. The structural system of a building is designed to hold a certain dead load — increases to this load could lead to deflection in the floor ("bending" of structural members) or, in the worst-case scenario, a structural failure. DIYers or contractors who want to install an earthen floor on a framed subfloor should check with a structural engineer to make sure there is adequate support for the additional weight of the earthen floor. Chapter 6 describes a simple test homeowners can do to check for stability and deflection in a framed subfloor before installing the floor.

Thermal mass

Thermal mass is the term used in building science to describe materials that help to moderate indoor temperature fluctuations in buildings by absorbing (and slowly releasing) heat. Materials that provide a lot of thermal mass are heavy and dense, like concrete or stone. (Interestingly, the material that can absorb the most heat per unit of volume is water.)

An earthen floor can absorb heat from the sun during the day and release it into the room at night.

(CREDIT: JOHN HUTTON)

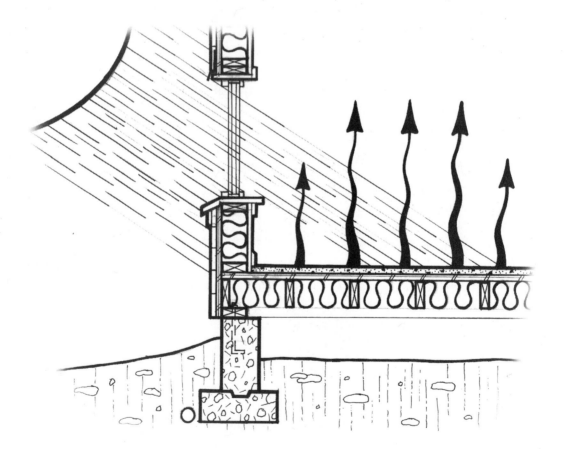

Bringing thermal mass into a home will affect its thermal performance. It will take longer to heat up from a totally cold, unheated starting point, but will hold heat longer. In climates that are warm and/or sunny during the day and cold at night, less heat input (from a fire or furnace) may be needed. To maximize the benefit of thermal mass, it should be used in conjunction with insulation (insulation outside, mass inside), and when possible located in front of south-facing windows, where it can receive direct solar energy.

Earthen floors are a good source of thermal mass in a building. While not as dense as concrete or stone, they still have the ability to store and release heat. Thicker floors provide more thermal mass. The question "How much mass is enough or too much?" is beyond the scope of this book, but generally speaking, bringing in thermal mass can benefit the thermal performance of any house, and an earthen floor by itself will never be "too much" thermal mass.

There are many texts available about passive solar design and strategies for using thermal mass, a few of which are listed in the Resources appendix.

Water beading up on a properly sealed earthen floor.
(Credit: James Thomson)

"Waterproofness" (or resistance to moisture and stains)

A common question about earthen floors is "What happens when they get wet?" When people think about a floor made of earth, they often imagine non-stabilized dirt, essentially like the ground outside. While earthen floors are made from the same materials, they are carefully finished with drying oils (and often wax) to create a durable, washable finish.

*Earthen floors can even be
installed in bathrooms!*

(Credit: Miri Stebivka)

Spilling a glass of water or wine on the floor will not damage it.

Excessive moisture though, especially if long-lasting and recurring, can be harmful. With prolonged exposure, water can seep into hairline cracks or at the edges and get in under the sealed surface. If the floor doesn't have a chance to dry, the material below can soften and shift, causing cracks or delamination of the oiled layer. Earthen floors are not recommended for areas that may receive frequent exposure to lots of water, such as outdoors, next to a pool or hot tub or in an open-shower bathroom. There are, however, lots of examples of earthen floors in kitchens and bathrooms that have lasted a long time, so these areas should not be ruled out.

Earthen floors can be stained by spilled oil from cooking. Usually these stains are not obtrusive; they look like a darker spot on the floor, tend to fade over time and usually vanish altogether when the floor is re-oiled or re-waxed (see Chapter 10). There have also been reports that pet urine, not immediately cleaned up, can discolor the surface of an earthen floor.

Toxicity

More and more people are becoming aware of the huge amounts of toxic chemicals that go into producing many conventional building products, and are justifiably concerned about bringing these poisons into their homes. Earthen floors contain mostly natural, non-toxic ingredients.

The main ingredients in earthen floors are things found in nature: sand, clay and natural fibers. The only materials with a potential for toxic ingredients are the pigments and the oils.

Natural pigments made from natural ochers and oxides are recommended, and can be harvested from local soil or purchased. Many commercially available pigments are produced with hazardous chemicals or harvested in a destructive way. See Chapter 3: Materials for more on the production and contents of pigments.

The drying oil and solvent used to finish an earthen floor pose the biggest risk of bringing hazardous chemicals into your home. Choose an oil blend with no chemicals or heavy metals and a natural solvent like citrus oil (see Resources appendix for brands and dealers). Even natural solvents contain naturally occurring VOCs (volatile organic compounds) that evaporate within about two days of application. The installer should wear a respirator with an organic vapor filter. Make sure the space is well ventilated and sealed off from any rooms that are being lived in. Once dry, the VOCs have all evaporated, and there is no known health risk to the occupants (see Chapter 3: Materials and Oil and Wax Safety appendix for more about solvents).

Many choose to finish their floors with a final coat of floor wax, much like the wax used on a wood floor. There are natural waxes available that do not contain any chemical ingredients (see Resources appendix).

Having a non-toxic floor is great for the homeowner and is also great for the floor installer (who may also be the homeowner). Most of the toxicity risks associated with building products are worst for the installers, who have to deal with the products while they are drying and off-gassing, and typically have to work with them over and over again on each job site.

Flammability

The bulk of a floor mixture is sand and clay, materials that simply won't burn. The fiber in the mix can burn in theory, but it is encased

On the toxicity of building products

It is only recently that homeowners have begun to ask the question "What's in that?" about the products their homes are built of. Stories of people becoming sick from carpets off-gassing, formaldehyde in insulation, chemicals in drywall and more have become more prevalent.

The solution seems easy: don't use products that have these nasty ingredients! But in reality this is more challenging than it sounds. First, there are few chemical-free options, and where there are options, the costs can put them out of reach for most. Second, it takes time to research the ingredients in particular product, and it is not always easy to understand what they are if they can be found. Most products do not include ingredient labels (like those on food products), though federal law does require manufacturers and importers of products containing hazardous chemicals to provide an MSDS, or Material Safety Data Sheet. While some MSDSs are useful guides, they may not offer a complete picture. Manufacturers often further obfuscate the ingredients of their product by listing them as simply "proprietary" or by giving chemical names that those of us without a chemistry degree won't understand. Finally, the science of what is and isn't hazardous is always evolving; consider that at one time asbestos was considered the best and safest fireproofing option. What will be the asbestos of the future?

Disposal Considerations (ANSI Section 13)

Waste disposal : Dispose in accordance with all applicable regulations. Avoid discharge to natural waters.

Regulatory Information (ANSI Section 15)

As of the date of this MSDS, all of the components in this product are listed (or are otherwise exempt from listing) on the TSCA inventory. This product has been classified in accordance with the hazard criteria of the CPR (controlled products regulations) and the MSDS contains all the information required by the CPR.

Physical Data (ANSI Sections 1, 9, and 14)

Product Code	Description	Wt. / Gal.	VOC gr. / ltr.	% Volatile by Volume	Flash Point	Boiling Range	HMIS	DOT, proper shipping name
1060-1200	vapor barrier primer sealer white	10.47	90.87	65.83	none	212-453	210	paint ** protect from freezing **

First page of an MSDS for a paint-on vapor barrier.

Ingredients   Product Codes with % by Weight (ANSI Section 2)

Chemical Name	Common Name	CAS. No.	1060-1200
ethanol, 2-(2-butoxyethoxy)-	diethylene glycol monobutyl ether	112-34-5	1-5
mica	mica	12001-26-2	10-20
titanium oxide	titanium dioxide	13463-67-7	5-10
propanoic acid, 2-methyl-, monoester with 2,2,4-trimethyl-1,3-pentanediol	texanol	25265-77-4	1-5
butanedioic acid, methylene-, polymer with 1,3-butadiene, ethenylbenzene and 2-methyl-2-propenoic acid	carboxy modified styrene butadiene polymer	52831-07-9	10-20
kieselguhr	diatomaceous earth, uncalcined	61790-53-2	1-5
water	water	7732-18-5	40-50

Chemical Hazard Data   (ANSI Sections 2, 8, 11, and 15)

Common Name	CAS. No.	ACGIH-TLV				OSHA-PEL				S.R. Std.	S2	S3	CC	H	M	N	I	O
		8-Hour TWA	STEL	C	S	8-Hour TWA	STEL	C	S									
diethylene glycol monobutyl ether	112-34-5	not est.	not est.	not est.	not est.	not est.	not est.	not est.	not est.	not est.	n	y	n	y	n	n	n	n
mica	12001-26-2	3 mg/m^3	not est.	not est.	not est.	3 mg/m^3	not est.	not est.	not est.	not est.	n	n	n	n	n	n	n	n
titanium dioxide	13463-67-7	10 mg/m^3	not est.	not est.	not est.	10 mg/m^3	not est.	not est.	not est.	not est.	n	n	n	n	n	y	y	n
texanol	25265-77-4	not est.	not est.	not est.	not est.	not est.	not est.	not est.	not est.	not est.	n	n	n	n	n	n	n	n
diatomaceous earth, uncalcined	61790-53-2	not est.	not est.	not est.	not est.	6 mg/m^3	not est.	not est.	not est.	not est.	n	n	n	n	n	n	n	n

Footnotes:

C = Ceiling - Concentration that should not be exceeded, even instantaneously.
S = Skin - Additional exposure, over and above airborn exposure, may result from skin absorption.
n/a = not applicable
not est. = not established

CC = CERCLA Chemical
ppm = parts per million
mg/m³ = milligrams per cubic meter
Sup Conf = SupplierConfidential
S2 = Sara Section 302 EHS
S3 = Sara Section 313 Chemical
S.R. Std. = Supplier Recommended Standard

H = Hazardous Air Pollutant,
M = Marine Pollutant
P = Pollutant,
S = Severe Pollutant
Carcinogenicity Listed By:
N = NTP, **I** = IARC, **O** = OSHA,
y = yes, **n** = no

in the non-flammable sand/clay mix. The sealing oils are combustible during their oxidation phase, but not explosive, and should be handled with care (see Oil and Wax Safety appendix).

Because of their non-flammability, earthen floors are good options around wood-burning stoves or in front of open fireplaces. However, it is very important to consider the materials in the subfloor. An earthen floor provides only a limited heat-shielding effect — heat can pass through it, and it is possible that a wooden subfloor below could burn. It would be wise to consult with a professional stove or fireplace installer.

A woodstove can sit directly on an earthen floor.
(Credit: Khaliqa Baki)

Odor

The floor mix itself has no odor once it dries, but the oil and wax do have a characteristic smell. As discussed above, the solvents in the oil and wax release some volatile organic compounds (VOCs) while drying. Once dry, the floor may continue to emit an odor, a nutty/piney smell, from the linseed oil. This smell is mild and will decline over time. How long the smell lingers will depend on the size of the floor, the temperature and ventilation of the space and how deeply the oil has penetrated into it. It is not uncommon to still be able to detect an odor after two or more years. Many people report that they like the smell.

Embodied energy

All manufactured products (and even many "natural" ones) contain "embodied energy."[4] This is the energy it takes to produce the product and deliver it to the building site. Cement, the active ingredient in concrete, has a very high embodied energy; it is made by cooking rocks (primarily limestone) at high temperatures for a long period of time, requiring lots of fuel. The finished product must then be transported from the mill to the job site, further adding to its embodied energy.

It takes very little energy to produce the raw ingredients of earthen floors. They may be dug by hand (giving an extremely low embodied energy) or by an excavator. An excavator runs on fossil fuels, but it can move a tremendous amount of material with a relatively small amount of fuel. A couple of gallons of diesel could be enough to dig and sift many cubic yards of dirt, enough for several homes.

The biggest energy cost for earthen floors is in the delivery of the materials. It is often possible to harvest some of the materials on or very near the site, but usually either sand or clay (and sometimes both) need to be imported. Depending on the size of the floor, this could require anywhere from a small pickup truck-sized load to a large dump truck-sized load (see Chapter 5 for guidance on calculating material quantities).

A diesel excavator at work.
(CREDIT: © MIKE KWOK | DREAMSTIME.COM)

Embodied Energy and Carbon Summary Adapted from Inventory of Carbon and Energy, University of Bath, 2006		
Materials	**Embodied Energy (MJ/Kg)**	**Embodied Carbon (KgCO$_2$/Kg)**
Carpet	74.4	3.97
Concrete for floor slabs	1.1	0.163
Portland Cement (sticky part of concrete)	4.8+-2	0.82
Rammed soil	0.45	0.024
Sand	0.1	0.0053
Steel	24	1.82

A 10-cubic-yard dump truck delivering materials.

(Credit: © Zorandim | Dreamstime.com)

"Even when produced by a machine, a finished earthen slab is estimated to have 90% lower embodied energy than finished concrete."[5]

Gasoline is still relatively cheap, so there may not be much financial difference between hauling materials five miles or fifty, but any serious effort to reduce environmental impact should carefully consider the environmental cost of using carbon-based fuel. Thus the shorter the distance the material has to travel, the better.

Color

Earthen floors can exhibit a wide array of rich colors (see photo in color section). The color is primarily determined by the natural clay color, which could be black, gray, purple, red, orange, yellow, green or brown. Other colors can be created by adding pigments to the mud mix or the sealing oil, or by applying a color wash.

Pigments and washes offer a lot of creative flexibility, but it should be understood that earthen floors are inherently dark and earth-toned. The finishing oils significantly darken the color of the dry mix, so even very light-colored raw materials will produce dark-colored floors.

Thickness

Earthen floors can vary from as little as ½" to over 3" thick. In addition to affecting the weight of the floor, floor thickness will affect ceiling height (by raising the floor level), thresholds at entryways and possibly even counter heights (if pouring a floor in a room that already has countertops installed).

Determining the desired thickness of the finished floor is one of the first decisions that must be made in any installation. See Chapter 6 for a discussion of thickness considerations.

Installation and drying times

Installing an earthen floor takes two to four weeks. This means the space is unusable for an extended period of time. The finish pour is ideally completed in one day, though if the floor is large or the installers are prone to taking nice long breaks, it is possible to leave a "wet edge" and come back to it the next day. Burnishing, oiling, waxing and buffing the wax take one day each as well. The bulk of the time, then, is for drying. Drying a wet floor usually takes a week; once oil is applied it needs a week to cure. If wax is used, it should be allowed to dry for another three to five days before the floor is buffed, and then be given another few days before it is ready for use.

Since drying is the most time-consuming part of the process, it is helpful to maximize drying conditions with heaters, air movement and ventilation, and humidity control. See Chapter 8 for more information about drying floors.

Cost

Natural building techniques have developed a reputation for being inexpensive, but this reputation is misleading. They can be inexpensive for someone who is willing to do the work themselves, from the harvesting and processing of the materials to the final installation. In this case, a person is exchanging time and effort for financial savings. Finding the right balance between "doing it yourself" and hiring someone else (or paying for premade materials) is the challenge for every builder and DIYer, regardless of the nature of the project.

One advantage of earthen floors is that it is possible to find nearly all the required materials for free or very little cost, if one is willing to put in the work. The main ingredients — sand, clay soil and fiber — can often be harvested onsite or nearby, or purchased at a very low cost. Processing these materials is time-consuming but simple work that can be done by a homeowner or a paid unskilled laborer. Alternatively, in some situations it may be worth buying a premixed bagged floor product that requires no material

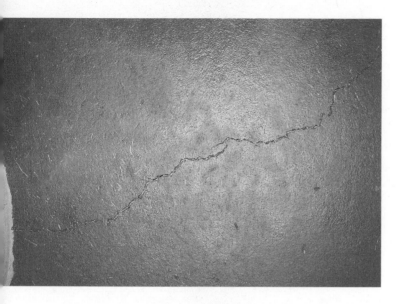

Cracks can happen and are not necessarily a problem!
(Credit: Sukita Reay Crimmel)

preparation labor (see Chapter 14 for more information).

The other ingredients — oil, wax and pigments — are more expensive and nearly impossible to find for free. Pigment and wax are optional, but the oil is a necessary cost to create a long-lasting floor.

Cracks

An earthen floor is a monolithic mass that can only be so large before it is prone to cracking. Cracks are a common occurrence in any type of continuous "slab" flooring system, such as earthen or concrete floors. They are caused by movement during or after drying, which is a result of settling, expansion with temperature swings, contraction during drying or dissimilar rates of drying. To put it bluntly, "cracks happen." A few small cracks are not necessarily a sign of weakness or a problem in the floor, and should be considered an aesthetic element that adds natural character and beauty to it. The installation instructions (chapters 8 and 9) offer guidance on minimizing cracks and how to treat them when they do occur.

SECTION 2

MATERIALS AND TOOLS

As with any building technique or project, it is important to understand the materials you'll be working with and the tools you'll need to get the job done right. Natural materials can not be purchased at a local building supplier; they must be identified, collected and processed. A novice builder will need guidance about where to find them and what to do with them. Similarly, gathering a good selection of proper tools ahead of time will save a lot of hassle down the road. (PHOTOS: LEFT: JAMES THOMSON, RIGHT: MIRI STEBIVKA)

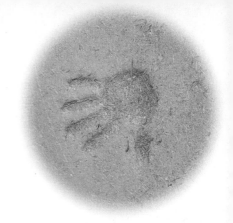

CHAPTER 3

Materials

ONE OF THE REASONS
PEOPLE LOVE EARTH-
EN FLOORS is because they are
mostly made of simple stuff that
anyone can understand, identify
and find themselves. The main
ingredients of sand, clay and
fiber evoke memories of playing
in the mud or at the beach as
a child. Some might still enjoy
making the occasional sandcastle or mud pie. These basic
"earth" elements are all around
us.

What is the earth
made of?

The earth underfoot is a mixture of many things. It is full of
organic matter (decomposing

Mud brings out the kid in all of us! (CREDIT: CRAIG KELLEY)

39

plant and animal bits), rocks and sands of various sizes, clay and, of course, living things like worms and bacteria. Different places on the planet have distinct soil contents, but the main ingredients are the same the world over.

The composition of the earth changes as you go down. The upper part, often referred to as "topsoil," contains most of the organic material in the soil. This is where the bulk of plant roots and living critters can be found.

Further down is the "subsoil," a layer that has little organic material and is a good place to look for clay-rich soil suitable for floors. Subsoils are a mix of sand, gravel, rock, silt and clay.

A properly made earthen floor is not just shovelfuls of earth from the yard spread on the floor, rather it is a carefully mixed and processed recipe with three main ingredients: sand, clay soil and fiber. Sand provides the structure and strength for the floor; it is essentially the building block. Clay is the binder that sticks everything together. Fiber provides tensile strength, which helps to minimize cracking, and an aesthetic touch. It may be possible to harvest some of these ingredients from the building site, but usually floor builders end up importing some or all of them from off-site.

A single gram of topsoil can contain from 40 million to more than 2 billion microbes.[6]

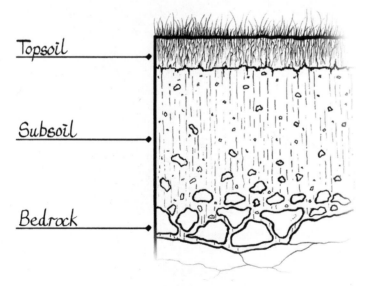

Typical cross section of the Earth's crust. (Credit: John Hutton)

Sand

Sand is a collection of very small rocks. It occurs naturally wherever larger rocks are broken down into smaller pieces, usually by water.

Rivers, glaciers, waves, rain: given enough time these forms of water will invariably erode big rocks into smaller ones. Rivers and waves tumble and crash against rocks, fracturing them into ever smaller and smaller pieces; glaciers grind away at mountaintops, reducing them to piles of rubble and dust; rainwater seeps into cracks in rocks and then freezes and expands, cracking large rocks into ever smaller pebbles.

The humble sand grain is the earthen floor's building block. Though small, it still maintains the compressive strength of its larger parent. An earthen floor is lots and lots of tiny sand grains stuck together (with clay) to create a strong, dense material that can survive years of use.

Sand grains come in various sizes. The largest grains are easy to see with the naked eye; anything from the smallest grains of beach sand to small pebbles fall into this category (larger grains are considered "gravel"). Keep grinding and the eventual result is silt, a dust-like substance that is the end product of millennia of geological activity. Silt is a common component of soils.

The best sand for floors contains a mix of grain sizes that

Silt

Silt is a common component of most soils, and will thus be present in most floor mixes. Dry silt looks and feels like dust; when wet it creates a squishy mud that can be confused with clay. But silt is not clay! The tests described later in the chapter will help differentiate clay-rich from silt-rich soil.

Some silt in a floor mix is a good thing; the small particles fill the spaces between the large ones, making for a dense and solid floor. Too much silt, however, can actually cause cracking. Keep the silt content to 30 percent or less in the final mix. This is a difficult thing to actually measure, because it's unlikely that the amount of silt present in the soil will be known (unless samples are sent to a lab for analysis, see Testing for clay content below). For most situations, a lab analysis is unnecessary; experimenting with the materials will provide sufficient guidance as to their suitability.

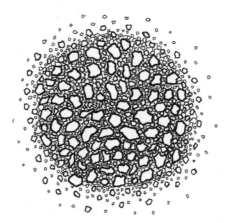

Close-up of sand, showing various grain sizes. (Credit: John Hutton)

Mason sand.

(Credit: James Thomson)

can pass through a ⅛" screen. Grains that are too large (>⅛") will create a mix that is hard to trowel smooth. A mix of grain sizes is beneficial because the small grains can fill the space between the larger ones, making for a more solid finished product. Sharp, angular grains are better than rounded ones for the way they lock and stack together. Beach sand does not make a good building sand, because the grains are uniformly sized and rounded from prolonged wave action.

No one is going to look at every grain of sand with a magnifying glass to determine if it is suitable for a floor mix. It is important to understand, though, what the "ideal" sand looks like and where it comes from, in order to use the best material available.

Sand can be harvested by hand or purchased. If harvesting by hand, a good place to look is along the banks of rivers and streams; sand in these areas tends to be broken down into small enough pieces, but has not been rounded by the continuous pounding of waves. Another good option is to look for decomposed granite, often referred to by its initials, DG. DG is the result of granite rock breaking down over time to create small grains. It is a great option for floors because it is sharp, variable in size and usually very clean of organic material. Harvested sand will likely have to be sifted (see Sifting, later in this chapter).

Harvesting sand is very labor intensive, so builders usually choose to purchase sand from a sand and gravel supplier (often called a "gravel yard" or "landscape supply yard"), where it can be bought in bulk relatively inexpensively. Because sand and gravel are heavy, it is not practical to transport them too far. For this reason, gravel yards that sell sand can be found in most towns.

Delivery may cost more than the material itself. Usually it costs as much to deliver ten cubic yards (the size of an average dump truck) as to deliver one cubic yard, so it is best to buy in bulk and have extra for other projects. Small quantities can be picked up with a pickup truck, but keep in mind that a standard-size pickup can only carry about half a yard of sand; more than that and the weight could put a serious strain on the suspension system. To calculate the quantities needed for any particular project, see Chapter 5.

Grades and varieties of sand available will vary from supplier to supplier. "Mason sand" is a common name for a fine-grained sand used for masonry that is suitable for finish floors. If a basecoat is being installed (see Chapter 6), a more coarse material can be used, for example material that could pass through a ¼" screen. It's best to visit the supply yard and look at the sand in person before it is delivered.

A handful of clay soil.

(CREDIT: MIRI STEBIVKA)

Concrete and masonry supply stores will sell mason sand in 50- to 100-pound bags.

Clay, or "clay soil"

Clay is the glue in the mix. Most naturally occurring subsoil clays can be used in earthen floor construction. Clay makes up a small percentage (by volume) of the total floor but plays a critical role in binding everything together.

Most people are familiar with the look and feel of clay, whether it is from childhood (or adulthood) pottery classes or from mucking around in the mud by

a lake or stream. Clay is very common in the Earth's surface, but not all muddy earth is true clay. Furthermore, there are a variety of different kinds of clays with different properties, and some are better suited for floors than others. Clay is made up of very small particles, similar in size to silt (see sidebar, Silt page 41), but clay molecules are not simply very fine particles of rock, they are tiny grains of aluminum silicates, a kind of mineral. Molecularly they are wide and flat in shape and have a slight negative charge that allows them to bond well with water. Dry clay expands when mixed with water. Some types will expand only a little, while other varieties (like bentonite) can expand three times or more. The volume change between wet and dry clay is referred to as its "shrink-swell capacity". When the water evaporates out of wet clay, shrinking occurs. This is what gives dry lake-beds their characteristic cracked appearance. Floors can crack because of shrinkage too, which is one reason to add sand to the mix (less clay=less material to shrink and crack), and to avoid highly expansive clay varieties.

Cracked dry earth is a indicator of clay content.
(CREDIT: SUKITA REAY CRIMMEL)

Sometimes clay minerals develop naturally in high concentrations, but most clay deposits are the result of sedimentary deposition; over time the grains of aluminum silicates wash downstream and concentrate in the bottom of lakes. Over several millennia, the clay mixes with silt, sand and organic material, creating a mix referred to as "clay soil." This clay-rich material can be found in many parts of the world, sitting at or just below the Earth's surface.

The ideal material for floors is rich in clay and low in silt, rocks and organic material.

Because of its unique properties, clay-rich soil is easy to recognize. Experienced earth builders can quickly identify suitable soil by touch, but novice builders will need some guidance to know what to look for.

Testing for clay content

An ideal soil has at least 20 percent clay and less than 30 percent silt. It's helpful to know approximately how much clay is present in the soil that will be used for the floor mix. The most accurate way

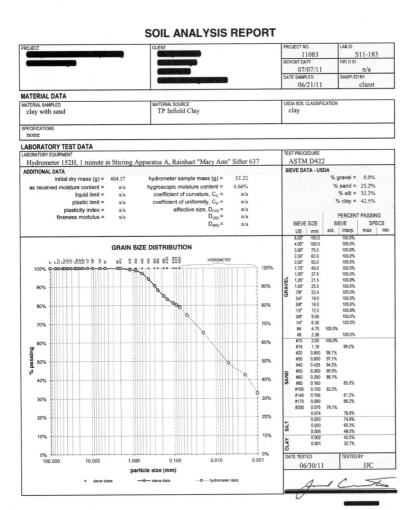

SOIL ANALYSIS REPORT

PROJECT	CLIENT	PROJECT NO.	LAB ID
▮▮▮	▮▮▮	11083	S11-183
		REPORT DATE	FIELD ID
		07/07/11	n/a
		DATE SAMPLED	SAMPLED BY
		06/21/11	client

MATERIAL DATA

MATERIAL SAMPLED	MATERIAL SOURCE	USDA SOIL CLASSIFICATION
clay with sand	TP Infield Clay	clay

SPECIFICATIONS
none

LABORATORY TEST DATA

LABORATORY EQUIPMENT	TEST PROCEDURE
Hydrometer 152H, 1 minute in Stirring Apparatus A, Rainhart "Mary Ann" Sifter 637	ASTM D422

ADDITIONAL DATA

initial dry mass (g) =	404.57	hydrometer sample mass (g) =	52.22
as received moisture content =	n/a	hygroscopic moisture content =	4.66%
liquid limit =	n/a	coefficient of curvature, C_c =	n/a
plastic limit =	n/a	coefficient of uniformity, C_u =	n/a
plasticity index =	n/a	effective size, $D_{(10)}$ =	n/a
fineness modulus =	n/a	$D_{(30)}$ =	n/a
		$D_{(60)}$ =	n/a

SIEVE DATA - USDA

% gravel =	0.0%
% sand =	25.2%
% silt =	32.2%
% clay =	42.5%

	SIEVE SIZE		PERCENT PASSING		SPECS	
	US	mm	act.	SIEVE interp.	max	min
GRAVEL	6.00"	150.0		100.0%		
	4.00"	100.0		100.0%		
	3.00"	75.0		100.0%		
	2.50"	63.0		100.0%		
	2.00"	50.0		100.0%		
	1.75"	45.0		100.0%		
	1.50"	37.5		100.0%		
	1.25"	31.5		100.0%		
	1.00"	25.0		100.0%		
	7/8"	22.4		100.0%		
	3/4"	19.0		100.0%		
	5/8"	16.0		100.0%		
	1/2"	12.5		100.0%		
	3/8"	9.50		100.0%		
	1/4"	6.30		100.0%		
	#4	4.75	100.0%			
	#8	2.36		100.0%		
SAND	#10	2.00	100.0%			
	#16	1.18		99.2%		
	#20	0.850		98.7%		
	#30	0.600		97.1%		
	#40	0.425		94.2%		
	#50	0.300		90.5%		
	#60	0.250		88.1%		
	#80	0.180		83.3%	85.0%	
	#100	0.150		83.3%		
	#140	0.106		81.2%		
	#170	0.090		80.2%		
	#200	0.075	79.1%			
		0.074		78.9%		
SILT		0.050		74.8%		
		0.020		65.3%		
		0.005		49.0%		
CLAY		0.002		42.5%		
		0.001		32.7%		

GRAIN SIZE DISTRIBUTION

HYDROMETER

% passing vs particle size (mm)

+ sieve sizes   —o— sieve data   ----o---- hydrometer data

DATE TESTED	TESTED BY
06/30/11	JJC

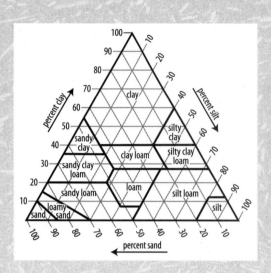

Soil type chart

This chart is a visual way to categorize the different types of soil. The types of soil best suited for earthen floors are:

Clay

Sandy clay

Sandy clay loam

Clay loam

Loam

to determine this is to send a soil sample to a soil lab for testing (or purchase a soil from a landscape supplier that has already been analyzed). The results will give the soil type name (see sidebar above) and the percentage and particle sizes of the sand, silt and clay present.

Most of the time a soil's composition does not need to be determined with such accuracy, and can be understood by simpler methods. The best way is to just start working with the material; experience is the best guide. Take a handful of soil, add water and mix it into a thick paste. If there is clay in the soil, the wet material will be sticky, malleable and slightly shiny. Silty soil is sometimes mistaken for clay, because wet silt has a similar texture and feel. But silt does not have the same elasticity as clay; it cannot be molded into forms. Try these tests to determine if there's enough clay for building:

Snake test

Roll a lump of wet soil into a small "snake" and wrap it around a finger. There should be little cracking in the snake.

Squeeze test

Form the clay into a ball and squeeze it in your fist. If it's rich in clay, the material will extrude out through the spaces between the fingers, creating ribbons of clay soil. Squeezing a wet ball of silt will not result in ribbons of clay but rather a sloppy mess.

Hairy arm test

Smear some very wet clay onto your harm and let dry, then try to peel it off. If the dried material tugs at the hairs on your arm (and hurts a little), then there is clay in the mix. If it rubs off without tugging the hairs, there is little clay present. *Note: This test only works for those with hairy arms!*

Shake test

Fill a glass jar about one-third with water and add loose soil until the jar is three-quarters full. Add a tiny amount of soap to help break the surface tension. Put the lid on and shake. Set the jar down and allow the contents to settle. The heaviest particles (sand and small pebbles) will settle to the bottom right away. Smaller grains of sand and then silt will settle

The snake test: this sample of clay-rich soil easily wraps around a finger without much cracking.
(Credit: Miri Stebivka)

Wet clay soil squeezed through fingers, forming "ribbons" of clay.
(Credit: Miri Stebivka)

Mix clay soil in water and shake (right). The different particle sizes will separate out in layers, showing their relative concentration in the soil (left).
(Credit: Miri Stebivka)

Soil science

Soil science is an incredibly complex field that even today does not receive the level of study and attention that it should. Leonardo da Vinci observed: "We know more about the movement of the celestial bodies than the soil underfoot." Five hundred years later the situation has not improved much. Digging into the details of soil is beyond the scope of this book; there are suggestions in the bibliography for further reading on this topic.

out next, taking up to fifteen minutes. Clay will settle out the slowest; in fact, the water could remain cloudy for hours or even days. If the water turns clear in the first fifteen minutes, this is a sign that there is not enough clay to build with. Once all the layers have settled, it is possible to observe and measure the relative quantity of each of the soil's components. Dividing the measured height of the clay layer (the top layer) by the total height of all the material in the jar will give the rough percentage of clay present. The same can be done for the silt and sand.

Once a suitable building soil has been identified, collect some for recipe testing. It may have to be sifted if it contains pebbles or rocks larger than ⅛" (See "Processing sand and clay," below).

Processing sand and clay

Self-harvested materials (and sometimes even purchased ones) will likely have to be sifted before use. For information on making screens for sifting, see Chapter 4. The screen size to use ("screen size" refers to the openings in the screen, not the size of the screen frame) will depend on the desired level of finish. For a basecoat (see sidebar, Chapter 6), ingredients can be coarsely sifted through a ¼" screen. If the material doesn't contain many particles larger than ¼", sifting can be skipped entirely. For finish coats, all the ingredients should be able to pass through a ⅛" screen. This makes for easier troweling and smoothing, and ensures a tight-grained surface. Purchased mason sand will likely not need to be sifted.

Sifting technique: For a small screen frame, lay the frame over a wheelbarrow or large barrel. Place some unsifted material on the screen and push it around over the screen surface by hand or with a piece of wood, allowing the sifted material to collect in the wheelbarrow. Do not force or "grate" the material through the screen! When the bulk of the smaller particles have sifted out, dump the leftover material and start with a fresh pile. An alternative technique is that two people can work together to shake the

Sukita:

People often ask whether they can use dry bagged clay from pottery stores. Most of my floor experience has been with native site soils, harvested from construction or other types of excavation. My few experiences using dry kaolin clay (a light-colored pottery-grade clay) with sand have raised questions about the strength of this mix, leading me to believe that the diversity of ingredients in common clay soil makes it superior to pure bagged clay.

Screening clay soil into a wheel barrow. (Credit: Miri Stebivka)

Two people sifting clay soil onto a tarp.
(Credit: James Thomson)

One person using a large angled screen to sift soil.
(Credit: Chris Foraker)

screen back and forth over a tarp. To more quickly sift a large amount of material, set up a large screen at an incline. Throw shovelfuls of material high up on the screen; as the material rolls down the inclined surface the smaller particles will fall through and the larger ones will end up in a pile at the base of the screen. Periodically shovel this pile away or move the screen.

It is always easiest to sift materials when they are dry. Wet materials will stick to the screen and clog it up, slowing down the sifting process. If possible, spread wet material out on a tarp to dry in the sun before sifting. Sometimes dry clay will need to be pulverized with a hand tamper or mattock before being sifted.

If the clay is already very wet, or pulverizing it before sifting is impractical, it can be soaked in water and mixed into a slurry (called "slip"), and then poured through a screen. This is a less desirable method, as it is impossible to know exactly how much clay is suspended in the slip. As long as the same batch is used for the entire floor mix, there shouldn't be a problem. Mix up the slip from time to time to make sure the clay is not settling to the bottom. Making more mix at a later time could cause difficulties; some water may have evaporated from the slip, changing the concentration of clay.

If the sand or clay soil is very rocky, do a "rough" sifting first through a larger-sized (½" to ¾") screen before using the finer (⅛") screen. This makes the sifting process more efficient.

Fiber

Fiber is the final main ingredient in an earthen floor. It provides tensile strength, which is important for long-term durability and to prevent cracking. For a fiber to be suitable for a floor it must meet certain criteria; it should be strong, short (<1" long) and contain little nutritional value for critters and mold. The most commonly used fiber is chopped straw. Straw is readily available in most parts of the world where there is grain production. If straw is unavailable, or a different aesthetic is desired, it is possible to use fibers from other plants, paper

Sifting wet soil is possible but messy and more work. (CREDIT: MIRI STEBIVKA)

Finely chopped straw is a great fiber for earthen floors.

(CREDIT: MIRI STEBIVKA)

**HAY IS FOR HORSES
STRAW IS FOR HOUSES**

CASBA   www.strawbuilding.or(

A bumper sticker from the California Straw Bale Association.

fiber or even manure. These options are discussed in Chapter 11.

What is straw?

Straw is the stalks left over from the production of grain. "Grain" refers to any of the cereal grains used for food: oats, wheat, barley, rye and rice all fall into this category. When grain is ready to be harvested, the seed heads are cut off, leaving the stalks behind in the field. These stalks are dried, cut and packed into bales. Modern agriculture has produced an incredible array of complex machinery that allows one person with a tractor to turn acres of grain into bales of straw in a single day. Of course in some places grain is still harvested by hand, with a scythe. Either way, the seeds are separated from the stalk and the stalks are "waste," so using straw is using a waste product. Straw should not be confused with hay, which contains dried grass, alfalfa and other nutrient-rich plants and is used to feed cows, horses and the like. Straw is not edible and is used for animal bedding, mulch and, of course, for building!

In addition to providing tensile strength, another benefit to using straw is the aesthetic touch it offers. The light straw fibers stand out against the dark floor, giving it a flecked, speckled look that many people love.

A bailing machine turns loose cut straw into tightly packed bales. (CREDIT: © MICHAEL SMITH | DREAMSTIME.COM)

One bale of straw is more than enough material for even a very large floor. Bales can be purchased at local farm-and-feed stores, landscape supply stores and, where grain is grown, directly from the farmers themselves. It will need to be

Straw fibers in an earthen floor. (CREDIT: JAMES THOMSON)

James demonstrates the proper use of safety gear while chopping straw. (CREDIT: MIRI STEBIVKA)

Seeds

A freshly installed earthen floor can take a week or more to dry. Any seeds in the mix may germinate and sprout out of the smooth finished floor. Yikes! Any sprouts that do pop out will die as the floor dries and can be removed before it is oiled.

chopped and sifted before it can be used in the floor mix.

Processing straw

Straw can be chopped into small pieces using an electric or gas powered leaf mulcher (see Chapter 4: Tools) and sifted through a ⅛" screen to remove seeds and straw nodes. Follow the safety precautions on the chopping machine. Wear safety goggles, ear protection and a particulate respirator, as straw chopping creates a lot of fine dust.

Cut open the straw bale, then separate out and loosen a few sections of it. Put loose straw into the chopper one handful at a time, being careful not to overload the motor. It is often possible to tell how hard the motor is working by listening to its pitch; as material is added, the pitch will fall as the motor slows down. Experiment with how much material can be added at a time to ensure the motor does not slow down too much.

The chopped material can then be collected in a barrel or on a tarp, depending on how the machine is set up. It may need to be chopped a second time to

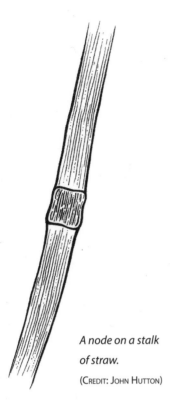

*A node on a stalk
of straw.*
(CREDIT: JOHN HUTTON)

ensure consistently short fibers.
Sift the chopped straw through
a ⅛" screen to remove any
leftover seeds, which can sprout
and cause a real headache.

The sifting also separates out
the larger straw "nodes," which
can catch on the trowel during
the final smoothing of the floor.
It's far easier to remove them
first than to try and deal with
them later. *(Note: For a basecoat pour, it is not necessary to sift the
straw.)*

*"Drumming" a screen to sift
straw.* (CREDIT: MIRI STEBIVKA)

Straw is a little more challenging to sift than sand or clay. One
technique that works well is to put the screen over a wheelbarrow or

barrel, place a few handfuls of straw on the screen, and then hit the screen with repeated firm and quick taps, as if playing a drum. This causes the straw on the screen to bounce up and down, and shakes the smaller strands through the screen. After a few minutes, dump what's left on the screen into a waste pile (this material can be used for a basecoat or as mulch in the garden) and repeat with a fresh pile of chopped straw.

Oil

The use of sealers for earthen floors has evolved over generations. Historically, and still in some places today, earthen floors were not stabilized or sealed — they were just earth. They were kept smooth and dust free by frequent tamping and moistening with water. Today, modern earthen floors use an oil-based sealer to provide durability and water resistance. Sealing an earthen floor involves saturating the top layer (up to ½" deep) with a sealant that transforms the raw earthen materials into a water-resistant, durable substance that will not simply dust away. Linseed oil is the most commonly used oil. It is derived from flax seed and was (and still is) used as a wood finish and a base for oil-based paints. The original linoleum was made from cork and cloth sealed with linseed oil. Raw untreated linseed oil is edible and according to some dietitians is a valuable nutritional supplement (flaxseed oil can be found at many health food stores).

Linseed oil is considered a "drying oil." Drying oils are oils that, through a chemical reaction with oxygen (oxidation), harden into strong polymer

Cross section of an oil-saturated earthen floor. The oil layer is the dark top layer, about ½" deep.

(Credit: Miri Stebivka)

chains. Some commonly used drying oils are linseed, tung, poppy seed, perilla and walnut oil. Raw untreated linseed oil oxidizes (dries) very slowly—it can be used as a sealer, but more commonly the raw oil is treated to speed up the drying process. Treated linseed oil is often referred to as "Boiled" linseed oil. Traditionally linseed oil was heated in a controlled process to purify the oil and promote oxidation. Today, boiled linseed oil sold at hardware stores and building supply stores is usually treated with chemical drying agents that contain heavy metals such as cobalt and manganese. These additives speed up the drying process, likely more than the traditional heat-treatment method. As with most chemicals, the greatest exposure risks are to those who manufacture these products or use them repeatedly over long periods of time.

Many people have used these kinds of chemically treated oils for their floors without incident, but because of the potential health risks associated with the drying chemicals, they are not recommended. There are a few commercially available linseed oils and linseed oil blends (which may also include tung oil, pine rosin,

Example of linseed oil found at hardware stores, usually treated with chemical dryers.
(CREDIT: JAMES THOMSON)

beeswax and/or natural citrus and pine solvents) that use a more traditional heat treatment to reduce drying time and don't contain chemicals or heavy metals. This type of oil is more expensive than the hardware store variety, but in the opinion of many builders, worth the additional cost. Typically linseed oil is thinned with a solvent to help it penetrate more deeply into the surface it's being applied on (wood or earth). Distributors of natural linseed oils, solvents and blends are listed in the Resources appendix.

Sukita:

When I first started doing floors, I was trained to apply the oil one coat at a time and allow each coat to dry a day or two before applying the next one. Also the first coat was 100% oil, the second coat 75% oil / 25% thinner, the third coat 50% oil / 50% thinner and the fourth coat 25% oil / 75% thinner. The logic was that the thinner would help the oil penetrate as the floor got more saturated with oil. Chad Tate, with MudCrafters in Colorado, shared with me the practice of pouring all the oil in one eight-hour period, the theory being that all the oil starts to oxidize at once. Further experimentation showed that it was not necessary to make further dilutions of the oil for increased penetration. The oil mix that I use now is diluted slightly (about 25%) with a natural thinner, and I have found no evidence that more dilute mixes improve penetration.

Solvents

Solvents are chemicals used to thin or dissolve a "solute" such as linseed oil. Many solvents are derived from petrochemicals (acetone, benzene), while others have more natural origins (ethanol, methanol). Two naturally occurring solvents are d-limonene and dipentene. These are both "food-grade" solvents (which does not mean you can eat them, but rather once they have evaporated you could eat off whatever they were applied to). The first is citrus oil, extracted by distillation from citrus peels; the second is a naturally occurring chemical found in a variety of softwood trees, plants and fruits.

Processing Oil (and solvent)

Mixing a heat-treated drying oil (e.g., boiled linseed oil) or drying oil blend with a solvent helps to thin the oil and promotes deeper penetration into the floor. There are oil-blends available that already contain solvent, or installers may choose to purchase the oils and solvents separately and mix them themselves. A ratio of three parts oil to one part solvent (75% oil, 25% solvent) makes a great sealer for earthen floors. One gallon of oil mix (purchased or mixed on-site) will cover 35 to 45 square feet of floor. The Resources appendix lists brands and suppliers of both products. Read the Oil and Wax Safety appendix before using oils or solvents.

It is possible to purchase raw, untreated linseed oil and treat it at home instead of buying a pre-treated "Boiled" linseed oil. This is not recommended, as the process is time consuming, inconsistent, and potentially dangerous, but for the committed DIYer or in situations where pre-treated oil is not available, it is an option. If you choose to take this on, be sure to research the process and proceed carefully; linseed oil is flammable, and over-heating will degrade the oil. Oil treatment techniques are beyond the scope of this book. A great book on oil finishes, *Shellac, Linseed Oil, & Paint,* provides more information on this subject and is listed in the resources section.

Wax

The application of a floor wax is the final step in the floor-making process. It's also an optional step; it adds some additional shine and water repellency but is not necessary. After the earthen floor has been sealed with oil and the oils have cured for at least a week, the wax can be applied (see Chapter 9: Finishing the Floor).

Floor waxes contain various kinds of wax, such as beeswax (produced by bees to form storage cells for honey and bee larvae) and/or carnauba wax (from the leaves of a palm tree native to Brazil). They also contain oil (often linseed) and a solvent. Most installers choose to purchase a wax; suggested brands are listed in the Resources appendix. Some have experimented with homemade wax blends; see Chapter 11 for further discussion.

Beeswax. (Credit: Miri Stebivka)

Pigments

Color pigments are not always used in earthen floor installations; most people want the "natural" look and choose to keep the natural color of the clay soil in the mix. But pigments can be used to liven up a room or coordinate with a desired color scheme.

Earthen floors are generally dark in color. This is because the

James:

My colleague Coenraad decided he wanted a light-colored earthen floor. He went out and bought white sand and light clay and made his mix with those materials. It dried to a gleaming white ... but that was before he oiled it. After he had applied multiple coats of oil, it had darkened down to brown ... the same color it would have been if he had used the native clay!

raw ingredients (clay, sand) are dark, and the sealing oil further darkens the mix. To alter the color requires a strong pigment that will show up in an already dark mix.

Pigments are what make color in our world. They are used in everything from makeup and house paints to clothing and fine art. Pigments may be naturally harvested, or synthetically produced in a factory or lab. Adding pigment to your floor introduces a potential for toxicity and negative environmental impacts that you should consider and attempt to mitigate.

Historically, naturally occurring pigments were mined and used with little consideration of their contents. Tests now reveal that some of these pigments may have contained naturally occurring concentrations of lead, mercury or asbestos. Most pigments found today are free of these dangerous compounds.

"Synthetic" or chemically produced pigments are made by subjecting combinations of naturally occurring minerals such as iron, silica and aluminum to chemical oxidation or a heat treatment. These are the same ingredients found in naturally occurring pigments, but processed by human hands to achieve the desired color. The production process often involves acids that can be hazardous to the health of the people making them, and the environment, if not properly handled.

Another consideration is the environmental impact of mining pigments. Copper, a common element in green and blue pigments, is mined in a very destructive way that leaves huge scars on the Earth and poisons soils and groundwater.

In some cases, there are sociopolitical problems related to the extraction and processing of the pigment. Lapis lazuli, a relatively rare semi-precious stone that can be processed into a blue pigment, has been mined in northern Afghanistan for more than three thousand years. The current political unrest in this area has made it less available. A synthesized version called "ultramarine" is now more common.

Whether the pigment occurs naturally or is created by a chemical process, it is a good idea to always ask the manufacturer for a MSDS for it to see what is in it and to avoid bringing hazardous elements inside.

Pigments extracted from animals and plants are rarely used in building; though they are often used to color fabrics. These pigments are typically not very strong in color and will likely fade in time. Examples include cochineal (from a bug), indigo (from a fermented tea plant) and Tyrian purple (from the mucus of a mussel). For an adventure in the history of pigment, read the book *Color: A Natural History of the Palette*, by Victoria Finlay.

Different pigments.

(CREDIT: MIRI STEBIVKA)

Sourcing and using pigments

It's possible to find a naturally occurring source of pigment, like a brightly colored clay deposit or an iron-oxide vein in the Earth, and harvest it by hand. But most of the time, pigments are purchased. The Resources appendix lists a few companies that sell natural pigments, but there are other more conventional options as well:

• Pigments used in concrete and masonry will work with earthen floors, and are widely available at building supply stores.

Sukita:

Adding titanium dioxide, a white pigment available at pottery supply stores, can bring the mix to a more neutral color to which colored pigments can then be added. I have started with a red clay soil, mixed in white titanium oxide and then mixed in a blue pigment to get a deep green! The red clay had warm yellowish undertones, which mixed with the blue makes the green.

• Pigments used in pottery are also suitable. It is important to get pure pigments; glazes are often mixed with other additives, which could include heavy metals. Pottery supply stores can be found in many towns and city centers, and of course online.

Most pigments come in powder form, though sometimes they are suspended in a liquid. It is important to wear a particulate respirator when working with pigment powder (or any fine-powdery material). The powders are usually sold in one- or five-pound bags. The amount needed for a floor will depend on the vibrancy of the pigment, and of course the size of the floor. Make tests with various amounts of pigments when determining the floor mix recipe (see Chapter 5). Pigment powder should be mixed with some water before adding it to the floor mix, as this helps ensure that all the pigment particles dissolve fully. Skipping this step can lead to streaking in the final mix as the undissolved pigment gets dragged by the trowel during installation. Some types of pigment dissolve more readily than others; examine the pigment solution after mixing with water. If there are still some undissolved particles, add a few drops of dish soap.

Pigment is also used in color washes and can be mixed with the finishing oil. See Chapter 11: Artistic Touches and Other Techniques for more on these options.

Miscellaneous materials

There are a few other materials you may need to install an earthen floor. These include:

Masking materials: Protecting walls, trim and other clean areas is important (more on this in Chapter 7). You will need painter's tape and paper or plastic masking film; find it at a local building supply or paint supply store.

Vapor barriers: Chapter 6 discusses the preparation of a subfloor for an earthen floor. In some cases, a vapor barrier may need to be added. Use 6 mm black polyethylene plastic for best results.

Vapor retarder: A vapor retarder is a membrane that is not completely impervious to water but helps to slow down its movement. These are used when installing an earthen floor over a plywood subfloor, to prevent the plywood from becoming saturated with water. The Resources appendix lists some options.

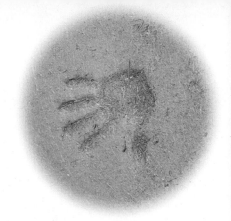

CHAPTER 4

Tools

THERE ARE A VARIETY OF WAYS TO MIX AND POUR EARTHEN FLOORS, and thus a variety of tools needed to do the work. This book suggests one or sometimes two methods that have been shown to work over years of experimentation and use. Builders often have their own preferences for ways to do certain jobs and the best tools to accomplish them with. The list here is only a recommendation, based on the authors' personal experience. Ultimately, tools that are critical to some may be unnecessary to others. Experience is the best guide.

Tools for gathering and preparing materials

Shovels

A shovel is an incredibly handy tool for digging holes to harvest clay, moving piles of dirt or transferring floor mix from a wheelbarrow to a bucket. A

Digging tools.

(CREDIT: MIRI STEBIVKA)

good shovel will last a lifetime. It's nice to have a pointy-tipped "round point" shovel for digging and a flat-edged "square point" shovel for scraping and moving piles of material.

Mattock (sometimes called a pick-ax)

This digging tool has a point at one end and a flat, axe-like blade at the other. It's great for harvesting soil and breaking up piles of dry packed earth.

James demonstrates an under-appreicated use for the wheel barrow. (CREDIT: MIRI STEBIVKA)

Wheelbarrow

Like a shovel, a wheelbarrow is an indispensable tool. Good for moving anything, it even makes a great chair! Buy a good one and use it forever.

Tarps

Woven polypropylene tarpaulins or tarps (often blue) have a variety of uses. They're good to put under piles of sifted sand or clay, and for covering piles in case of rain. They're available at any hardware store. A source of recycled tarps is the material used to wrap lumber for transport to a lumber yard. Ask a local lumber yard if they have any extra "lumber wrap."

Buckets

One can never have enough five-gallon plastic buckets. Seriously! Fortunately they are often available for free.

Good sources are grocery stores with bulk-food departments and restaurants (lots of food products come in five-gallon buckets). Collect at least twenty.

Straw-chopper

There are a few ways to chop straw. The most common method is to use a barrel leaf mulcher (see photo in Chapter 3), which is essentially an electric weed whacker inside a plastic barrel. Straw is added at the top, it falls onto whirling plastic strings (like in a weed whacker) and shorter pieces fall out the bottom. Note that the strings will have to be changed frequently, as the straw is a lot tougher than the leaves these machines are designed to chop. Warning: these machines are extremely noisy.

Another effective chopping tool is a small gas-powered chipper–shredder. These machines are intended to turn small-diameter sticks into wood chips, but they work quite well for straw. They are better than barrel leaf mulchers in that they have much more powerful engines, so they can work through more material more quickly, and

Buckets and tarps.
(Credit: Miri Stebivka)

Small gas-powered chipper-shredder.

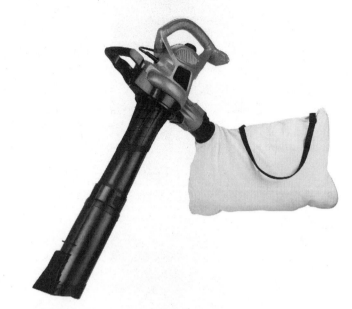

Electric leaf blower also will suck and chop straw.

Using a weed whacker in a trash barrel to chop straw. (Credit: James Thomson)

metal blades that don't need to be replaced. The disadvantages are that they are expensive (though often available for rent from a local shop) and heavy. Double chopping (rechopping material that has been passed through the machine already) may also be required.

Other options: Many electric leaf blowers also have a vacuum/chopping function. They will even collect the chopped material in a removable bag. These tools work slowly and don't chop as finely as other options, so double or triple chopping will be necessary.

A final option is to use a regular weed whacker in a trash barrel. Create a lid that fits around the weed whacker arm by cutting a slit out of an old trash lid; alternatively you can use a piece of heavy-duty fabric (burlap or canvas), fill the barrel about halfway with loose straw, stick in the weed whacker, put on the cover and engage the motor. This option is a bit awkward and produces an inconsistent product, but may be preferable to buying a piece of equipment that you will only use once.

Whichever straw-chopping tool you use, make sure to follow the safety precautions that come with it. Wear safety glasses, ear protection and a particulate respirator.

Screens and screening

Screens are used to sift the raw ingredients to remove large pieces and make a more consistent mix for a finished floor. They can be made with scrap wood and hardware cloth. Hardware cloth is metal screen material, sold in rolls (or sometimes by the foot) and available in several different-sized openings. The most common size is ⅛", but ¼" and even ½" screens can be handy.

Make the frame out of wood and attach the screen securely to it. Small screens (roughly 2' x 3') can be used over a wheelbarrow or trash barrel, while a larger screen can be set up at an angle over a tarp.

Tools for making the mix
Buckets, shovels, wheelbarrow (as above)
Measuring spoons, cups and buckets

Variety of screen frames. (Credit: Miri Stebivka)

Different sizes of hardware cloth. (Credit: Miri Stebivka)

For making test samples, a set of small measuring cups is very useful. At a minimum, you will need a few one-quart yogurt containers and a one-cup measurer. If pigments are being used, you will also need measuring spoons for making test samples.

When it comes to producing the final mix, use a bucket with tape or marks on the side to indicate sand and clay proportions (see photo page 98).

Tools for mixing

For small floors, it is easy to mix the materials in a wheelbarrow with a hoe, or on a tarp with your feet. See Chapter 5: Making the Floor Mix for more instructions.

For a large floor, mechanical mixing assistance is recommended. A mortar mixer is a great mixing tool; this machine has a fixed barrel with rotating paddles inside, as opposed to a cement mixer (a rotating barrel with paddles fixed to the inside). This type of mixer gives a consistent mix and is worth its price for the labor it saves.

A gas-powered mortar mixer.

(CREDIT: MIRI STEBIVKA)

They are available for rent at most tool rental shops. Do not operate any heavy equipment without proper training and knowledge of its safe operation!

A small cement mixer can also work, but will take longer to mix and won't be able to mix as large a batch. The mixes will also need to be wetter to mix effectively.

Tools for pouring

Buckets, shovels, wheelbarrows (as above)

Levels

Making a floor level is very important, and requires familiarity with the use of levels. The minimum requirement would be a 2' and 4' spirit (or "bubble") level. Longer ones (6' or even 8') can be very useful too, if available. It is a good idea to keep mud and carpentry tools separate and not use an expensive carpentry level on a muddy floor.

A laser level is a very useful tool that is highly recommended for improving efficiency and quality. There are many brands available; find one that has a bright beam and can be attached to a wall with a bracket.

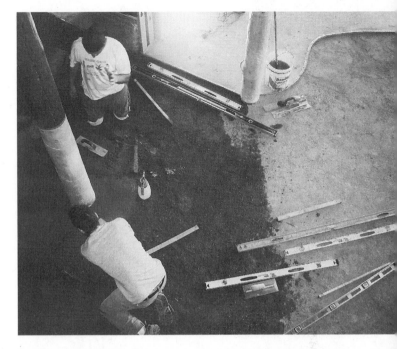

Levels in use. (CREDIT: SUKITA REAY CRIMMEL)

Laser level and tools for proper installation. (CREDIT: MIRI STEBIVKA)

Screed rails and boards ready for use. (CREDIT: JAMES THOMSON)

Floats and trowels for pouring. (CREDIT: MIRI STEBIVKA)

Using a laser level during the installation process is described in Chapter 7.

Screed boards and rails

Screeding is a technique well-known to concrete workers, and can be used to make a flat and level earthen floor, too. Screed rails are laid down parallel to each other on the subfloor, about two feet apart. The mix is dumped between the rails and then a screed board (a 2x4) is dragged across the rails using a side-to-side motion to spread and level out the mix. Screed rails can be made by ripping larger boards down to the desired thickness of the floor and cutting them to about four feet in length.

Trowels and floats

Wooden floats: Wooden floats are the first tool used to spread and level the floor. They are light, long and stiff, and help to reduce the undulations that are an inevitable result of hand-troweling. They can be handmade from a strong lightweight wood like cedar, or purchased at a mason supply store.

Steel trowels: These come in a range of shapes and sizes. A small selection of different ones is recommended. The most commonly used model is a stiff rectangular steel trowel, 3" x 12". This is the tool for smoothing the surface once the mix has been spread with the wooden float, and for doing the final burnishing. Also handy is a 7-inch or 12-inch pool trowel. This is a more flexible steel trowel with rounded corners, designed (unsurprisingly) to trowel the curved inside of cement pools. This type of trowel may be easier for the novice troweler to use, as the rounded corners are less likely to leave trowel marks. However, the more flexible pool trowel will leave more undulations in the surface of the floor; this is not a "problem," but a flat surface can only be achieved by using a rigid square trowel. A small, delicate "Japanese trowel" is very useful for tricky spots, though expensive (see Smoothing, Chapter 8, and Resources appendix for distributors).

Pool trowels are less likely to leave trowel marks, but can leave more undulations. (CREDIT: MIRI STEBIVKA)

Tools for burnishing
Trowel

For the burnishing stage, the stiff rectangular steel trowel works best.

Rigid foam used to get out on a still-damp floor to burnish it.
(CREDIT: JAMES THOMSON)

Stainless steel floor pans and other burnishing tools. (Credit: Miri Stebivka)

Floor pads

During this stage, the burnishers need to walk out onto the still-wet floor. To do this without damaging the floor, they require pads or pans to disperse their weight. These two options work in different ways:

Option 1: 1"–2" rigid foam insulation pads, cut into squares large enough for someone to kneel on comfortably (2' x 3' or 3' x 3'). Try taping the edges of the pads so insulation bits do not come off while in use and get stuck in the floor. Make two per installer.

Option 2: Stainless steel floor pans. These are used in the concrete floor industry, and work extremely well on earthen floors. The user slides them along the floor like skis.

A pump-action spray mister

This is used for misting down the floor with water if it has gotten too dry to burnish effectively. Use a two-quart mister; anything larger will be hard to move around. Make sure to clean it well when done. Sand grains can easily find their way into the pump mechanism and will quickly damage it.

Sponges

Large masonry sponges are useful for cleaning up spills or splatter, and for adding additional moisture to an area if needed.

Tools for drying

Unless warm, dry and breezy weather is guaranteed during the drying process, some mechanical assistance is strongly recommended to ensure quick and uniform drying.

Fans

Fans are critical to assist with quick drying. Use fans with high "cfm" (cubic feet per minute) ratings, and that can be moved around and directed at particular spots on the floor.

Dehumidifiers

If the weather outside is cool and/or humid, a dehumidifier may be necessary. An industrial, high-capacity dehumidifier (available at tool rental shops) works best, not only because it pulls more moisture out of the air but also because it comes

Pump-action spray mister.
(CREDIT: MIRI STEBIVKA)

Selection of fans. (CREDIT: MIRI STEBIVKA)

Industrial and home-style types of dehumidifiers.

(Credit: Miri Stebivka)

with a drain tube that can be run to a drain or out of the house. A household-grade dehumidifier will certainly help, but they are less powerful and the reservoir will have to be manually drained. Select a model that can take 60 to 150 pints of water out of the air per day.

Oil application tools

There are several ways to apply oil. *(Note: Anyone working with oil must be familiar with how to safely use, clean and dispose of oily brushes, rollers and rags. Read the Oil and Wax Safety appendix for details.)*

Brushes

A 3-inch brush for oil-based paints is needed for oiling the edges of the floor.

Rollers and roller grids

For larger floors, a paint roller with a long pole is very handy. Use half-inch (or higher) nap. Consider investing in a "roller grid," a

screen that fits inside a five-gallon bucket. Roll the roller back
and forth on the grid to remove
excess oil.

Rags

Rags are good for applying oil
and for cleaning up areas that
might have been over-saturat-
ed. Use cotton rags that do not
leave behind much lint.

Sprayers

There are pump-sprayers available for applying oil. Because of
solvents in the oils, it is important to get a sprayer with seals
that work with solvents, such
as Viton seals (see Resources
appendix for details); otherwise
the solvents may destroy the
seals. This option is primarily for those who are pouring
many floors or a very large floor.
Prescreen oil to remove any
small particles that could clog
up the spray nozzle.

Buffing/sanding tools

An optional step (described
in Chapter 9) calls for lightly
sanding the floor after oiling.
This helps create a smooth
final surface. The sanding can
be done with a small handheld

*The tools used to
oil an earthen floor.*
(CREDIT: MIRI STEBIVKA)

*A sprayer that is
used to spray oil.*
(CREDIT: MIRI STEBIVKA)

Large industrial floor buffer.

(Credit: Sukita Reay Crimmel)

Larger square floor sander.

(Credit: James Thomson)

orbital sander, an angle grinder, a large industrial floor buffer or a large square floor sander. The smaller tools are easier to manage for the novice and are suitable for small floors. The larger tools are better suited to large installations, and should not be handled by anyone unfamiliar with their use. All are typically available at local tool rental shops.

There are often two types of buffers available: a high-speed and a low-speed version, and different sizes and shapes. The high-speed model is actually easier to handle, but harder to find. The lower-speed tool is very hard to control. The square floor-sanding machine is a good compromise. Do not attempt to operate any of these tools without proper training.

The tools described above are designed to work with different types of sanding and buffing pads and disks. The preferred pad is a green scrubby one, manufactured specifically for floor finishing; it resembles the green scrubby pads common in kitchens but is one-inch thick and shaped to fit both large round buffers and square

Using handheld grinder with sanding pad.
(Credit: James Thomson)

Buffing pads, full-sized and custom-cut for smaller tools.
(Credit: James Thomson)

floor sanders. These pads can be purchased at tool rental stores, janitorial supply stores and, sometimes, flooring supply stores. If they are not available, 80-grit sandpaper will work, but is much more aggressive and will leave significant scratches on the surface (these will disappear with the wax application).

Handheld orbital sanders typically have a "hook and loop" pad that allows different sanding disks to be quickly attached and removed. A similar style of pad can be purchased for an angle grinder. A green scrubby pad will also stick to this surface; custom make one by cutting an appropriately sized disk out of a larger pad.

Wax application tools

Waxing the floor is another optional step, described in Chapter 9. It is the final step in the floor installation process.

Brush

Use a 3-inch oil-paint brush to wax around the edges. It can be the same brush used for the oil.

Lambswool wax applicator or rags

An 18-inch sheep's wool applicator with a long pole is the easiest and most effective tool for applying wax. Terry cloth rags or a regular sponge are also good options.

Wax buffer

After the wax has dried, it must be buffed to even it out and remove excess material. Buff with the same tool that was used for sanding (described above), but this time using a white scrubby pad, which is somewhat softer and less abrasive than the green pad. These pads are available wherever the green pads are sold, and can be cut down to size to fit a handheld sander or angle grinder.

Tacker and utility knife.

(Credit: Miri Stebivka)

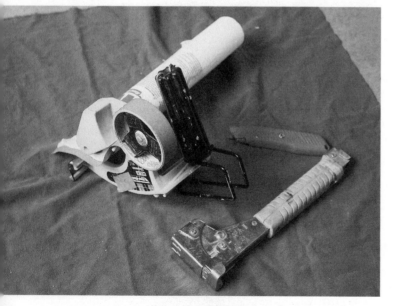

Tools for room and floor preparation

Protecting the walls and trim in the room is an important step before pouring begins. There are a few key tools that will make this job easier.

Masking paper (or plastic) dispenser

A very handy tool for masking the walls in preparation for the pour.

Hammer tacker with staples

For applying vapor retarder, if needed (see Chapter 6).

Hammer, pry bar, nail puller

For removing trim.

Safety equipment

Eye protection

Safety glasses should be worn at all times! Get a comfortable pair and get used to wearing them.

Work gloves

Indispensable gear for any work project. There are many different kinds, though some seem better suited for mud work than others.

Woven gloves with rubberized fingers and palms: These thin gloves offer less protection than thick leather gloves but allow for more

Common safety gear.
(CREDIT: MIRI STEBIVKA)

sensitivity and dexterity. Floor installations don't involve many sharp objects, so they are suitable.

Water- (and oil-) proof gloves: Some people like to use waterproof gloves for pouring and troweling, to keep their hands warm and dry. Choose a durable pair with a woven lining rather than the cheap dishwashing variety. These gloves are strongly recommended for the oiling and waxing steps.

Ear protection

Use ear protection when operating a mortar mixer or other loud machines or tools!

Respirator

When working with (or creating) fine particles, it is recommended you wear a particulate respirator. The white dust masks do not work well; invest in a half-face respirator with NIOSH-approved P100 filters. Most oils and waxes contain VOCs, so use a respirator with organic vapor cartridges during this step.

Knee protection

Whether they're troweling the wet mix, burnishing the surface or applying oil and wax, floor installers spend a lot of time on their knees. Knee protection is critical for comfort and to minimize joint problems down the road. There are several options: a wide array of strap-on kneepads is available for just this purpose. Sometimes these pads can be a liability when working with wet mix, which can collect between the pad and the knees, making the pads messy and less effective. A better option is to use a small foam kneepad that can be moved from spot to spot. There are a few varieties of these available too, from very cheap foam pads to a gel-filled version that is expensive but worth it.

Tool maintenance

Many tools are expensive, so it's worth taking the time to maintain them well so that they last a long time. Follow any maintenance guidelines listed in the tool instructions.

Because earthen floor work is inherently dirty, plan extra time at the end of each workday to clean tools. The general rule is to thoroughly wash and allow to dry all tools that get muddy. Use a stiff plastic-bristled brush for cleaning if necessary. It's wise to rinse out buckets and allow them to dry before stacking them together for storage. Buckets stacked when they were dirty or wet can be extremely hard to separate!

Power tools can be wiped with a damp cloth (unplug them first). Occasionally blowing out dust and chopped straw bits with pressurized air will keep the tools running longer.

Any tools used for oil or wax should be cleaned with a solvent, such as paint thinner or citrus solvent. Often people choose to leave brushes used for oil (and oil-based paints) soaking in solvent until their next use.

Rust spots on steel trowels.

(Credit: James Thomson)

Trowels and floats are the mainstay of a floor installer's tool collection, and some of them are quite expensive. They should be well-maintained to provide many years of use. If steel trowels get rusty, they will not do as good a job at smoothing the floor. Thoroughly clean trowels and floats after use, and allow them to dry. If any trowels have wooden handles, consider oiling them with a wood oil (like linseed or tung oils) for additional protection. Store trowels in a place where they won't get bent or nicked by other tools. For extra protection, oil the metal. Any kind of oil is suitable, though there are specially formulated tool oils available.

If the surface of a trowel does rust, it can be gently refinished using 220-grit wet/dry sandpaper. Wet the sandpaper and the trowel and rub off any imperfections.

SECTION 3

PREPARING FOR THE POUR

The materials have been gathered, the tools are ready, the spot for the floor is selected; now it's time to get to work on installing your floor! The first step is to develop a recipe for the mix. After that, the subfloor needs to be prepared, and the room needs to be prepared for the installation. All this needs to happen before a single bucket of mud hits the floor. Be patient! Skipping ahead without proper preparation will ultimately lead to a disappointing finish or more work down the road.

(PHOTOS: LEFT: MIRI STEBIVKA, RIGHT: JAMES THOMSON)

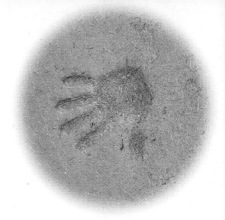

CHAPTER 5

Making the Floor Mix

Making the mix

MAKING A GOOD MIX can seem daunting to the novice builder. This chapter covers in detail the recommended steps and
processes for determining a recipe for and producing a finish floor
mix. If installation includes a basecoat pour (see sidebar in Chapter
6), the process is the same except that the batch for the basecoat
need not be finely sifted or processed. It is critical that any mix be
thoroughly tested before being installed; testing is the only way to
ensure a quality floor.

Finding the right recipe

The ingredients: The basic ingredients (discussed in Chapter 3) are
sand, clay (soil) and fiber. Collect and process enough for several
rounds of tests. The materials should be from the same location
(and processed in the same way) as the materials you will use in the
final floor.

Required tools: A few buckets, measuring containers, wood float,
a metal trowel and a place to spread the tests out to dry; a piece of
plywood will work, or a concrete slab.

Start by making several batches of sand and clay soil mixed together in different ratios. Keep in mind that earthen floors are mostly sand. The final ratio will depend a lot on the amount of clay in your clay soil; if it is very clay-rich, you will need a lot of sand to compensate.

Try starting with three batches in the following ratios:

Sand : clay
2 : 1
3 : 1
4 : 1

Each test batch should be about a half-gallon, so use a one-pint (two-cup) or one-quart (four-cup, the size of a standard yogurt container) container as a measure. For each test, mix the dry ingredients in a bucket, then add a little bit of water and mix again. Continue adding water until the mix reaches a thick "cake batter" consistency. Then spread it on the drying surface with a wood float and smooth it over with a steel trowel. The test patch should be about ¾" thick and at least 18" in diameter; bigger is better. If the patch is too small, any shrinking that may occur will not result in cracking, as the whole patch will shrink uniformly. It's important to know if the mix will shrink and crack, so make the patch big enough that this will show. Carefully label the test patch with its

Testing different ratios of sand to clay. The test patch on the left has the highest clay content, the one on the right has the lowest.

(Credit: James Thomson)

sand-to-clay ratio. Continue this process for each of the other test ratios, and let the test patches dry.

What to look for

Once the test patches are dry, look for cracks, brush your hand across the surface and scrape at it with a thumbnail.

Cracking: Cracks are caused by the shrinkage that occurs when clay dries. The best case is a test that has few or no cracks. If a test patch has a lot of cracking, the clay content is too high, so more sand should be added.

Dusting: Brush the surface with a hand and observe if sand grains come loose. It will be possible to feel them, see them or even hear them falling off. A small amount of dust and sand coming loose is normal, but too much dusting is a sign that more clay needs to be added.

Scratching: The test patch has not been stabilized with oil so it can be scratched with a fingernail, but it should feel hard and resistant. The material should not crumble or break away easily. A surface that is too fragile indicates low clay content; try a higher clay-to-sand ratio.

In most situations, more than one sample may pass the "touch test." If this is the case, look for other features, such as appearance: what does the surface look like? Are grains of sand visible? Does it have a tight, smooth surface? Make additional tests that further refine the ratios. For example, if both the 2:1 and 3:1 samples seem sufficient, try a 2½:1 or maybe even a 3½:1 test. Ultimately, if there is no discernible difference between two tests, use the mix with the higher sand percentage. This will help minimize the possibility of cracking.

Once you have determined the sand-to-clay ratio, make new test mixes using fiber. The amount of fiber will be measured as a percentage of the total volume of dry mix. For example, if the mix has three parts sand, one part clay and ½ part fiber, this would be 12.5% fiber (4 parts sand/clay, ½ part fiber = (0.5/4)100 = 12.5).

To make the fiber tests, start by mixing about two gallons of just the sand–clay mix (in the ratio determined above). Next, scoop out a quart of dry mix and transfer it to a clean bucket. Add water and mix to a cake batter consistency, then add ¼ cup of fiber and mix again, adding more water if necessary. Dump the test out on a flat surface, trowel it out and let it dry. Repeat this process, using ½ cup, ¾ cup and a full cup of fiber.

Here's an example to clarify the process:

During the dry mix testing, three parts sand to one part clay is determined to be the best ratio.

Using a 1-quart measure, measure 6 quarts of sand and 2 quarts of clay into a bucket and mix thoroughly (do not add any water). Scoop out 1 quart of this dry mix into a separate bucket, add water and mix to cake batter consistency. Sprinkle in ¼ cup of fiber and mix again, adding more water if necessary to bring it to a stiff batter-like consistency. Spread and smooth out on a flat surface, at least 18" in diameter and about ¾" thick, just like before. Record the percentage of fiber in this sample and label it. This sample contains 3 cups of sand, 1 cup of clay, and ¼ cup of fiber. This makes for 4 cups of dry mix (4 cups sand/clay), so ¼ of fiber would be 6.25% of the sand/clay portion. Another way to think of this is in parts: 12 parts sand, 4 parts clay and 1 part fiber.

Repeat this testing process with ½ cup, ¾ cup and a full cup of fiber. The ratios and percentages for each of these tests is as follows:

Mixing fibers

It's not uncommon to mix two (or more) types of fiber into a floor mix. Some fibers are shorter, like those from paper and manure, and make for a mix that is easier to work with, while longer straw fibers provide more tensile strength. A combination of short and long fibers makes for a good compromise. Adding a percentage of screened manure to the fiber content is a well-known practice. For more on alternative fibers, see Chapter 11: Artistic Touches and Other Techniques.

¼ cup fiber: 3 cups sand, 1 cup clay, ¼ cup fiber =
   4¼ cups total; 6.25% fiber (12:4:1)

½ cup fiber: 3 cups sand, 1 cup clay, ½ cup fiber:
   4½ cups total; 12.5% fiber (12:4:2, or 6:2:1)

¾ cup fiber: 3 cups sand, 1 cup clay, ¾ cup fiber:
   4¾ cups total; 18.75% fiber (12:4:3)

1 cup fiber: 3 cups sand, 1 cup clay, 1 cup fiber:
   5 cups total; 25% fiber (12:4:4, or 3:1:1)

Allow the test patches to dry and examine the results.

What to look for

Workability: The fiber in your mix alters the workability; more fiber makes it harder to spread and smooth out, and fiber strands could poke up through the mix.

Aesthetics: Fiber changes the aesthetic of the finished surface, a subjective characteristic. The person who will be living with the floor should be the one to decide what kind of look they like.

Strength: Fiber increases tensile strength and reduces cracking; too little fiber in the mix will be of little help in this capacity.

Testing for straw content, from lowest to highest (left to right). (Credit: James Thomson)

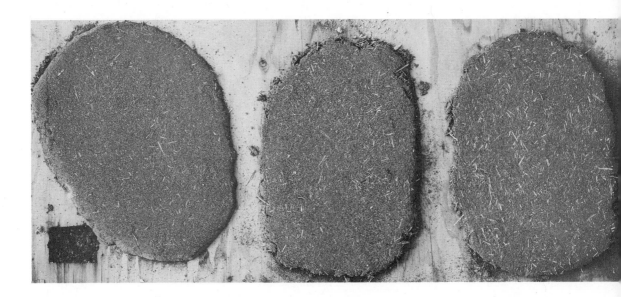

There is some flexibility here in how much fiber to use, though there are practical limits on the upper and lower thresholds. Too little fiber will do little to increase the strength and durability of your floor; too much will be difficult to work with and (in extreme cases) could make the floor soft and brittle. Tests should always be the guide for what looks and works the best.

If you are unsure about how much fiber to use, select the highest percentage that still offers good workability. A common range is 10%–20%.

Adding color

If you want to change the natural color of the mix, you can add pigment directly to it. Changes in color can also be achieved by adding a color wash or using pigmented oil; see Chapter 11 for more on these techniques. It's important to make tests with different pigment concentrations so you can find the color you desire. Use measuring spoons to add measured amounts of pigment to measured amounts of the dry sand, clay and straw (one teaspoon per cup of dry mix, for example), add water, mix well and spread it on a surface to dry. Make sure to record the amount of pigment in each test! Measuring by weight is the most accurate, but if this is not possible, be sure each scoop is packed the same to ensure uniform ratios. (Pigment should not exceed 5 percent of the volume of the total mix; more may result in cracking or dusting issues with the floor. Experimentation is encouraged, make plenty of tests.) When samples dry, apply oil to them to get an accurate representation of what the finished floor will look like. Once you've achieved the desired color, some math will be required to determine how much pigment to add to each batch in the final recipe, using larger measuring containers.

Final testing

Once you've determined a final recipe, you should do a final round of testing that will replicate everything you will do when installing the floor (burnishing, oiling, buffing, waxing). Read through the installation instructions in chapters 8 and 9 and decide which steps to include. For this final test, make a larger test patch, at least three feet square. This will give a more accurate representation of what it will be like to work with the material, and what the finished floor will look like. Once you've completed all the testing, you will have a solid recipe that has been shown to work, and a good amount of experience working with it. It is now just a question of "scaling" up to a full-size floor.

All this testing may seem like a lot of work, but it is a critical step, especially for beginners. With experience, the testing process goes much more quickly: it is easier to choose appropriate ratios of sand and clay, and the total number of test patches can be reduced. The final test can be eliminated as understanding of how burnishing and oiling affects the finished product increases.

Determining quantities

Now you need to calculate the quantities of raw materials you will need to make enough mix for your floor. The first step is to figure out the total amount of mix needed. This is a simple geometry problem: multiply the surface area of the floor by the thickness of the pour.

(Note: This means the thickness of the pour needs to be determined before mixing can begin. For help determining thickness, see Chapter 6.)

Example: Installing a floor in a 12- by 20-foot living room, at ¾" thick.

1) Determine the square footage of the floor: 12' x 20' = 240 ft².
2) Convert the thickness in inches to thickness in feet; ¾" = 0.75"; 0.75/12 = .0625".
3) Multiply the square footage by the thickness; 240 ft² x .0625 ft = 15 cubic feet (ft³).

This gives the total volume of the finished floor in cubic feet. It would seem that the total volume could be multiplied by the percentage of each ingredient to get the total volume for each ingredient, but experience has shown that this amount would be insufficient. The reason for this is that the clay consists of such fine grains it nearly dissolves in water; the microscopic grains fill the spaces between the sand grains, adding very little to the final volume of the mix. Similarly, dry straw reduces significantly in volume when wet and has a negligible effect on the final volume of the mix. To calculate the amount of material you will need, pretend that the mix is 100 percent sand, then add an additional 20 percent to the volume to make up for loss due to spillage and shrinkage.

The example continues to see how this works in practice.

Finished floor volume = 15 ft³. Add 20% to this (3 ft³) to get the total volume of sand needed = 15 ft³ x 1.2 = 18 ft³.

The recipe is 3 parts sand, 1 part clay and 15% of the sand/clay volume in fiber. Divide the amount of sand by 3 to get the amount of clay needed:

18ft³/3 = 6 ft³ of clay

The authors made a little bet...

You would think that if you took a bucket of sand and added a half a bucket of clay, you'd end up with a bucket and a half of mix, right? While this might be true when the materials are dry, adding water changes this equation....

Experience has shown that water dissolves the clay and fills the spaces between the sand grains, so that the final volume is less than the sum of the volumes of the two dry materials. But how much less? James theorized that the volume of clay plus sand (plus water) would still be more than the volume of sand alone, while

Sukita was certain that volume of the clay and water would not add to the final volume of the mix at all. And thus an experiment was designed, with measured amounts of sand and clay, and a bet ensued....

The result? The sand and clay were mixed together, water was added and the final volume of the mix was taken. Surprisingly, even with the addition of water and clay, the final volume of the mix was the same as the original volume of just the sand. Sukita won the bet, and James bought the first round of drinks.

For fiber, take 15% of the sand/clay volume = 18 ft³ + 6 ft³ = 24 ft³; 15 % of 24 = 3.6 ft³.

Cubic feet is an unfamiliar volume for most people, but it is standard when ordering from a sand and gravel supplier — in fact, they will require materials to be ordered in cubic yards (1 cubic yard is 3 cubic feet on each side, so 3'x3'x3' = 27 cubic feet). 18 ft³ of sand is 0.67 cubic yards (18ft³/27ft³); this happens to be ⅔ of a cubic yard. It is not cost-effective for gravel suppliers to deliver small quantities (less than 1 cubic yard, maybe more). Some gravel suppliers do have split trucks that could deliver bark mulch and gravel for other projects at the same time, to spread out the cost of delivery. Ask neighbors or other contractors working nearby if they need any materials, or get extra for future projects.

Split dump truck.

(Credit: James Thomson)

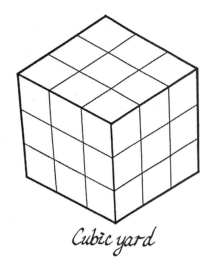

Cubic yard

(CREDIT: JOHN HUTTON)

Cubic foot

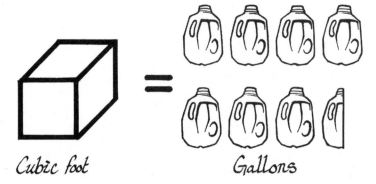

Cubic foot = Gallons

(CREDIT: JOHN HUTTON)

For harvesting material and mixing the final mix, gallons are a handy unit of measure. There are 7.5 gallons in a cubic foot.

Thus,

Sand: 18 ft^3 x 7.5 gal/ft^3 = 135 gallons
(27 5-gallon buckets)

Clay: 6 ft^3 x 7.5 gal/ft^3 = 45 gallons
(9 5-gallon buckets)

Fiber: 3.6 ft^3 x 7.5 gal/ft^3 = 27 gallons
(5.4 5-gallon buckets)

Pigment: If pigment is being added, the total amount must be calculated as well, based on the recipe determined earlier.

This gives the quantity of each raw ingredient needed for a 240 square-foot floor. There is one final ingredient, of course, and that is water. Make sure to have plenty of it available! Quantities needed will be discussed in the next section.

Mixing the mix

Earthen floor mixes should be made in batches of a manageable size. "Manageable," of course, is a relative term. The size of each batch will depend on the equipment available to mix it, and how strong the (human) mixers are feeling.

As discussed in Chapter 4, the mix can be done mechanically (with a mortar mixer) or by hand mixing (in a wheelbarrow). It's also possible (and fun and messy) to mix it on a tarp on the ground. Make this decision based on the size of the floor and the resources available (money, people, tools).

Caution: The mixing process involves lifting heavy material, dealing with fine particulates and potentially using a motorized mortar mixer. Operators must be well trained in the use of the mortar mixer and in proper body mechanics to prevent injury. Ear and eye protection and a particulate respirator are recommended.

Mixing in a mortar mixer

Mortar mixers will save a lot of effort and are recommended for larger pours or when working with a small crew. They are noisy and potentially dangerous, so make sure the operator is familiar with how they work. (Mortar mixers are good at churning and integrating the sand with the clay and straw, and are recommended over cement mixers. It is possible to use a cement mixer, but the batch size will need to be kept smaller and the mix will likely have to be wetter to ensure the ingredients get thoroughly mixed.)

Using a mortar mixer to make the floor mix.

(CREDIT: JAMES THOMSON)

First, determine how large each batch will be. Mortar mixers come in different sizes (usually measured in cubic feet). They are designed to mix concrete mortar, which is much thinner and easier to mix than earthen floor mix. If the batch becomes too large, the mixer will have a hard time turning and may stall. Generally, keep the batch volume to less than half of the mixer's total volume. Consider also that you will have to move each batch from the mixer to the floor. Ideally one batch would fit in a wheelbarrow,

though it is possible to split a single batch into two separate wheelbarrows. For an 8-cubic-feet (or 60-gallon) mixer, start by mixing about 25 gallons at a time, including all dry ingredients and the water, and adjust if necessary.

For the recipe determined above, divide up the ingredients as follows:

15 gallons sand
5 gallons clay
3 gallons fiber
~4 gallons of water (about 1 gallon per 5 gallons
    of final mix; start with less and add more until
    the proper consistency is achieved)

Measuring mark on a bucket for mixing.

(Credit: James Thomson)

This adds up to 27 gallons, but the clay, straw and water will fill into the sand, not contributing to the final volume, resulting in about 15 gallons of mix (see sidebar page 94). It is wise to select a mixing recipe that breaks down into convenient units, rather than fractions of gallons.

Designate a bucket (or several) as a measuring bucket by making a line on the side with a permanent marker or tape to mark the appropriate fill line.

Start by adding water. The total quantity of water added will depend on the size of the batch and the recipe. Add it slowly, a little at a time; it's easy to add more later but impossible to take it out.

If you're mixing a lot of batches, it's worth keeping track of how much water you put in the first batch to make the process of mixing subsequent batches more efficient. Use a quart or half-gallon container to measure in some water to start, then keep track of how much more gets added during the mixing process. For the next batch, start with the full amount used in the first batch minus 10 percent; the piles of sand and clay often have varying moisture levels in different areas, so some batches may need less water. Again, it's easy to add more. The person doing the mixing will get to know the look, sound and feel of the perfect batch. Listen to feedback from the trowelers and adjust batches as necessary.

Fire up the mixer and engage the paddles. If pigment is being used, add that next and let it mix with the water for a couple of minutes, to give it time to fully dissolve (see the section on

James:

Mortar mixers do save a lot of time, but sometimes saving time is not our only goal. We like to enjoy our work, and to stay healthy. I always encourage people to think carefully before reaching for a power tool or machine to do a job they could do with hand tools. It is easy to slip into the "faster is better" mind-set, but ultimately there is something wonderful about an engine-free work site. Hand mixing is quiet, meditative, a good workout, and there's little risk of getting injured. The sound of a mixer can be deafening, and really changes the feel of a job site. Mixers also give off exhaust, which is unpleasant to be around; are expensive; and can be dangerous, especially if you are unfamiliar with operating them.

I have often found that so-called time-saving devices, like power tools, heavy machines and so on, can actually set you back much further than doing the job by hand. A wrong move with a backhoe, for example, could knock over a finished wall that took weeks to build, or a large tree that took decades to grow. And because they are so powerful, power tools and machinery also cause much more serious accidents if something goes wrong. The risk of a serious injury, which could take weeks or months to recover from, should always be balanced with the potential time savings of using a machine. Of course, the building schedule (and my energy level) don't always allow for it, but if I can do it the old-fashioned way, I will.

pigments in Chapter 3: Materials). Then add the clay and let it
mix for at least a minute (usually the time it takes to carry the next
bucket over and pour it in is sufficient). Next add the sand. If the
proportions of water and dry materials are correct, the sand should
mix in easily to create a wet slurry. If the mix is too dry at this point,
the engine of the mixer will struggle and the paddles will slow
down (and possibly stop altogether). Add more water if this occurs.
The fiber will absorb quite a lot of moisture, so at this point the mix
can be wetter than its final consistency.

Finally, add the fiber. The mixer will work harder as the fiber
absorbs water and the mix stiffens up. Once all the fiber is mixed in,
the mix should be at the proper consistency. More water can always
be added if necessary.

Mixing in a wheelbarrow

Two people mixing floor mix in a wheelbarrow.
(Credit: Miri Stebivka)

Hand-mixed batches will be smaller than those made in a mortar
mixer. First calculate the quantities for the dry ingredients. Start

by trying a total batch size of about 15 gallons, with the following recipe for our example mix:

12 gallons sand
3 gallons clay
2¼ gallons fiber
~3 gallons water

Mix the sand and clay together first, *without any water*. Use a hoe (or two people and two hoes, standing at opposite ends of the wheelbarrow) to thoroughly mix the dry ingredients together. (This is more effective when working by hand than mixing the water and clay together first, because wet clay is so sticky that it can clump up and be difficult to mix. The mortar mixer doesn't have this problem because it's powerful and fast, but for mere mortals, it can be quite exhausting!) Next add the water, a little bit at a time, mixing throughout to bring the mix to a wet batter consistency. (If using pigment, premix the pigment with water and add it now.) As before, add water slowly so as not to add too much. Finally, add the fiber, and more water if necessary, to bring the batch to the proper consistency.

Mixing on a tarp on the ground

(For those who have mixed cob, this method will be familiar.)

This method works best with two people. Spread a polypropylene tarp (6'x8' or 8'x10') on the ground. The order is similar to mixing in a wheelbarrow. Dump the measured amount of dry sand and clay onto the tarp (shoot for a total batch size of 15–20 gallons, the same or a little larger than a wheelbarrow-sized batch). Use the tarp like a mixer: "toss" the dry materials back and forth, with each person holding two corners of the tarp and working together to mix the dry sand and clay. Once they're well mixed, pile up the ingredients in the center of the tarp and make a hole like a volcano crater. Pour some water (with pigment, if used) into the hole and mix again, this time stomping on the pile (with boots, or bare feet for extra fun!) to aid in the mixing process. Use the tarp

Mixing floor mix with bare feet on a tarp. (CREDIT: MIRI STEBIVKA)

to flip the material onto itself, like an omelette, and stomp on it again. Finally, add the fiber and mix again in the same way. Add more water if necessary until the mix reaches the proper consistency. Use the tarp to lift the finished batch into a wheelbarrow for easy transport. Be careful, it will be heavy! Don't try to lift it alone.

Mixing on a tarp is a lot of fun, and a great way to get people involved, but it can be messy. Because the mix is wetter than a standard cob mix, there is a risk that some will spill off the sides of the tarp during mixing. The mixing site will get dirty, and if too much material is lost off the tarp, it could alter the final ratios of ingredients.

Whatever the mixing method, it is important to note that piles of sand and clay often have different moisture contents in the middle than on the outside. The day's humidity will also affect the mix. Learning the feel of a good mix is one of the joys of working with earthen materials.

What's the right consistency?

There is no universal agreement among earthen floor aficionados about the "perfect" consistency; ultimately it is based on the preference of the installer. A wet mix is easier to move around and integrate but takes longer to dry and has more chances of cracking; a dry mix is easier to get flat but is harder to move around and may "stretch" more during installation, which could cause cracking. In cool weather or humid climates, it is best to keep the mix as dry as possible to reduce drying time.

There are lots of metaphors that could be used to describe consistency: cooked oatmeal, stiff cake batter, wet concrete. The testing process should offer some clues as to what the final consistency should be, but ultimately experience is the best guide. During the mixing process,

check the consistency periodically by feeling the material (turn off the mixer before reaching inside) and watching the way it moves as it is being mixed (in a mixer, it should rise up with the paddles and fall off at the top; in a wheelbarrow, it should hold its shape as it is moved around with a hoe without being too dry). Try this test:

Take an empty pint or quart container, fill it with the mix and then dump it out onto a flat surface, like making a sandcastle at the beach. If the mix settles down to half its height or less right away, it is too wet; if it keeps the same shape with no settling, it is too dry. It's easy to add more water during the mixing, just spray it in with a hose. If the mix is too wet, small amounts of dry ingredients can be added, in the proper ratios. Ideally each batch will have the same consistency (and water content), to ensure consistent drying.

Slump test showing an appropriate amount of "slumping" for a floor mix.
(Credit: James Thomson)

CHAPTER 6

The Subfloor

FOR EASE OF DISCUSSION, "subfloor" is defined in this book as any stable floor a finish coat of earthen floor will sit on. The criteria for building subfloors for earthen floors are similar to those for subfloors for any other kind of flooring. The main considerations are the additional weight of the earthen floor and the moisture present during installation.

Subfloors for earthen floors can be designed in a variety of ways and constructed out of a range of materials, including concrete, wood and compacted gravel (roadbase). For new construction, subfloors should be built according to current building code specifications, which will likely involve the inclusion of insulation and a vapor barrier. Subfloors that rest directly on the ground (called "on-grade" subfloors), such as concrete slabs or compacted gravel, can easily support the weight of an earthen floor. Wooden subfloors, built with beams, joists and plywood, are usually strong enough too, but in some cases, extra reinforcing may be necessary. Consult with a building professional to determine current requirements and whether additional reinforcement is needed.

One of the first decisions to make before installing an earthen floor is determining its thickness. The floor's overall weight will

105

Basecoats

On occasion, an earthen floor is poured in two (or more) layers, with a rough "basecoat" and a fine finish coat. Installing a basecoat is not common, but there are some situations where it may be desired:

1. To raise the floor level for design reasons. If the thresholds or transitions to adjacent flooring types would be easier to deal with if the finish floor level was a couple of inches higher, use a basecoat to add the additional thickness under the finish coat.

2. When using radiant heat. If the heat system design calls for more than an 1½ inches of material above the tubes or electric elements, pour a basecoat up to 2" or 3" thick to cover the tubes. See Chapter 13 for more on radiant heat.

3. To bring more thermal mass into the room. See Chapter 2 for a discussion of thermal mass. This is not commonly needed, as a 1"–2" earthen floor provides a lot of thermal mass.

4. To level out a non-level subfloor. If the subfloor is not level it is easier to make the ☞

A basecoat being poured over hydronic heating tubes. (Credit: Sukita Reay Crimmel)

correction with a basecoat and then pour a finish layer on top.

5. To provide additional stabilization. If working on a loose subfloor (e.g., an old concrete pad that has broken and cracked in places) or adding rigid foam insulation on top of a slab or compacted grade, a basecoat will add additional strength and stability.

   Note: This only applies when working on an on-grade subfloor of concrete or gravel. An unstable wood-framed subfloor will not benefit from the addition of the extra weight of a basecoat; instead additional structural support is needed.

For new construction, or during major remodeling, the basecoat should be poured early in the building process, to provide a level and stable surface to work on while building the rest of the house. Once dry, it is important to cover basecoat with plywood, to keep the dust down and reduce the amount of material that is rubbed off by foot traffic.

Also note that it is important to sweep and vacuum the loose material and wet down the dry base before pouring the finish.

depend on its thickness, and weight is a major consideration when installing on a framed subfloor (see discussion of weight in Chapter 2). Floor thickness also affects thresholds, headroom, trim and transitions to areas with different types of flooring.

Earthen floor thickness

Generally speaking, a floor should be as thin as possible while still providing sufficient strength and structure. Gathering, mixing and moving the raw materials is heavy and time-consuming work. A thinner floor means less work and faster drying, as well as less weight. But adequate thickness is needed to provide good solidity, and a thicker floor can help bring more thermal mass into a room, level out a non-level subfloor or integrate with a radiant heat system.

The recommended range for a finish earthen layer is ½" to 2". A pour as thin as ½" is possible over very stable subfloors (like a concrete pad or earthen basecoat). Over wood framing, at least ¾" is recommended. This is because a wooden subfloor will have more flex and bounce than an on-grade subfloor, and a thinner floor (less than ¾") could crack over time (see Chapter 12 for a story about a ½" floor over wood framing). Most floors fall into the ¾" to 1" range. This seems to be the sweet spot, a thickness that allows for good structure and mass without being too heavy or so thick that drying takes excessively long. Floors with hydronic radiant heat or where extra thermal mass is desired may be two inches thick or more. Thicker floors are possible but should be applied in multiple layers, with a rough basecoat (or two) and a fine finish layer. This helps to minimize problems that may arise when drying very thick layers, such as mold growth, cracking and sprouting.

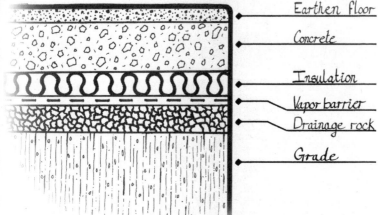

Cross section of an earthen floor on a concrete slab.

(Credit: John Hutton)

Earthen floor
Concrete
Insulation
Vapor barrier
Drainage rock
Grade

Types of subfloors and their suitability for earthen floors

Concrete slab

Concrete slabs are common floor systems for conventional homes. They are typically strong enough to support the weight of an earthen floor of almost any thickness. If the slab is free of cracks, the finish floor can be poured directly on it with little preparation. Sometimes the old concrete slab is in very poor structural shape, with lots of cracks or loose sections. There may be control joints (cut or scored lines, see Chapter 7) larger than ¼". There are a few

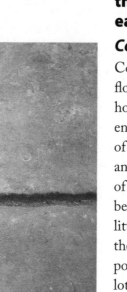

A large crack in a concrete subfloor that requires mitigation.

(Credit: James Thomson)

techniques to employ to make sure these cracks do not transfer up to the finished floor:

1) Pour a thick basecoat layer with plenty of chopped straw to create a solid base for the finish coat.
2) Separate the finish coat from the pad with a membrane. Use a vapor barrier, roofing paper, building felt or even ¼" foam to create a "break" between the cracked pad and the finish floor. Make sure to have enough material on top of the membrane to weigh it down and eliminate weak spots, 1½" at least.
3) Imbed a reinforcing "grid" into the floor. Examples include fiberglass stucco lath and loose-weave landscape cloth, often made of burlap, jute or hemp (see Chapter 12 for more about using burlap).

Compacted gravel (roadbase)

A compacted gravel subfloor is a good option for ground-level floors in new construction. This technique creates a solid, well-drained

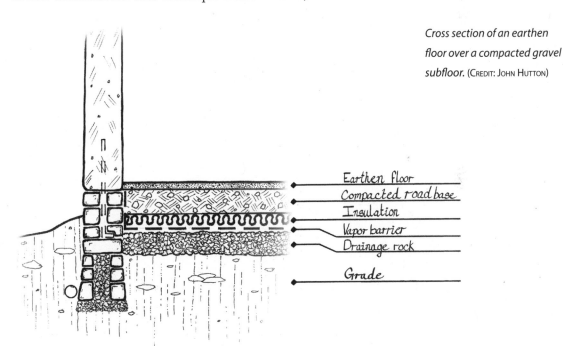

Cross section of an earthen floor over a compacted gravel subfloor. (Credit: John Hutton)

Earthen floor
Compacted road base
Insulation
Vapor barrier
Drainage rock
Grade

subfloor that minimizes the use of concrete. It is little used in conventional construction but is popular in the natural building field because it requires no cement, a high embodied energy product. See the Building a Compacted Gravel Subfloor appendix for guidance. Because the surface of a compacted gravel subfloor may include some loose gravel, a finish earthen layer of at least ¾" is recommended.

Framed wood subfloor

This is the standard type of floor in most conventionally built houses. The main considerations for pouring over wooden subfloors are whether the floor can support the weight and protecting the wood against the moisture in the earthen floor mix until it has a chance to dry.

Weight considerations: Most wood-framed subfloors, including second floors, built to recent code specifications should accommodate

Cross section of an earthen floor on a conventional wood-framed subfloor.
(Credit: John Hutton)

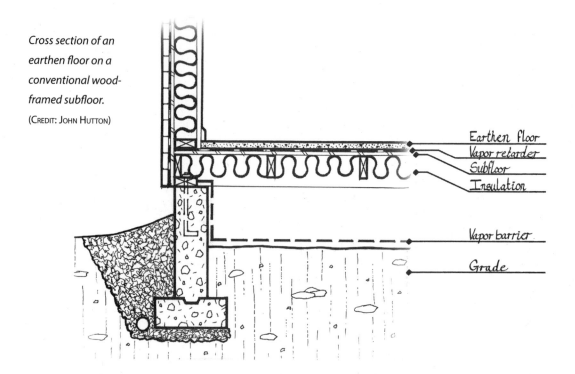

Earthen floor
Vapor retarder
Subfloor
Insulation

Vapor barrier

Grade

the weight of two inches of earthen floor. The best option is to consult a structural engineer to make sure your subfloor can support the weight. If this is not an option, it is possible to perform a simple deflection test. Deflection is the "give" or "bounce" in the floor. A floor should have a deflection value similar to what is needed for tile: L/360 or less. L/360 is the length of the unsupported span in inches ("L") divided by 360. For example, if the span is 10 feet, 10' x 12' = 120" divided by 360 = ⅓", so there should be no more than ⅓ of an inch of deflection.

Deflection test

Stretch a piece of string across the middle of the room, in the same direction that the floor joists run. Secure the string at both ends so it is parallel to the floor and a few inches above the floor level. Measure the length of the string in inches; this is the span, or L. Divide the span by 360. This number gives the "allowable" deflection for a floor that span, or L/360.

Then measure (in inches) the distance of the string from the floor at the walls; it should be the same at both ends. Apply a heavy weight to the middle of the room (a piano, five or more adults), then measure the distance from the string to the floor at this point. Subtract the height of the string measured at the wall from the height measured in the middle of the string while weight is present to get the total deflection. If the measured deflection is less than L/360, then the subfloor should be able to comfortably support an earthen floor. If the measured deflection exceeds this amount, additional support may be required.

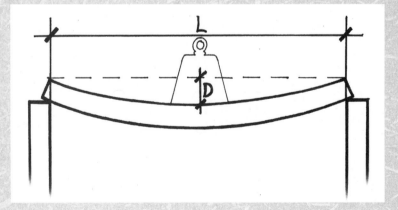

Illustration of deflection.
(Credit: John Hutton)

Moisture considerations: Protect the plywood from becoming saturated with water while installing and working with the wet earthen mix. Before the pour, affix a vapor retarder to the subfloor with a hammer tacker or tape. A vapor retarder is a membrane that helps to protect the wooden subfloor from getting wet during installation (which could cause it to warp). Tar paper and waxed craft paper are both examples of vapor retarders. See Resources appendix for product listings. Allow a three-inch overlap and tape the seams to prevent the vapor retarder from shifting during installation.

Elements of a subfloor

In addition to being structurally sound, a subfloor must protect the finished floor from moisture in the ground with a vapor barrier and a drainage layer, and provide insulation for efficiency. For an "on-grade" subfloor (concrete or compacted gravel),

Installing a vapor retarder over plywood subfloor. (Credit: James Thomson)

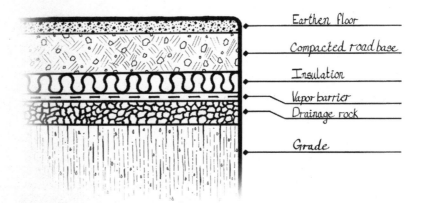

Cross section showing placement of insulation and vapor barrier in a compacted gravel subfloor.

(Credit: John Hutton)

Earthen floor
Compacted road base
Insulation
Vapor barrier
Drainage rock
Grade

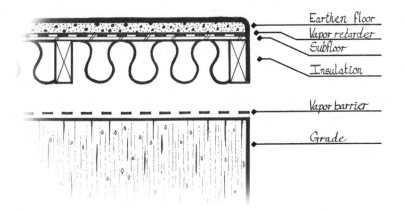

*Placement of vapor barrier
in a wood-framed subfloor.*
(Credit: John Hutton)

the vapor barrier should be placed under the insulation and above the drainage layer. Framed wood subfloors do not sit directly on the ground, so the vapor barrier and insulation can be handled differently. If there is a crawlspace, there should be a vapor barrier on top of the bare ground (under the joists), to protect the wood framing from rot.

Drainage

Every on-grade subfloor should have a drainage layer. This is usually a layer of drainage rock, and in very wet soils may also include drainage pipes and/or a French drain. An important role of the drainage layer is to provide a capillary break, so that moisture in the ground cannot easily wick up into the upper layers of the subfloor. See Building a Compacted Gravel Subfloor appendix for more information, and consult a building professional if needed.

Vapor barrier

The ground contains moisture, which evaporates into water vapor. This vapor can migrate from the ground below up into the house; condense back into liquid water and cause increased humidity, mold growth and, in extreme cases, rot.

A vapor barrier is a waterproof membrane that blocks the movement of water vapor. Most building codes require the use of one in new construction; 6 mm black polyethylene plastic sheeting

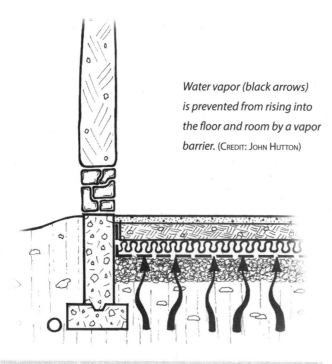

Water vapor (black arrows) is prevented from rising into the floor and room by a vapor barrier. (Credit: John Hutton)

Why use a vapor barrier?

Earthen floors are permeable and will allow water vapor to pass through them. This vapor can condense on the floor under objects placed there and can eventually lead to discoloration or mold growth. Many earthen floors have no vapor barriers, and though the owners may find the buildup of moisture under objects to be annoying, it is certainly a manageable problem. Keep large objects like beds off the floor and take out the rugs to dry every so often. Mold growth can be removed with a mild hydrogen peroxide solution and typically does not recur if the surface stays dry. Building a vapor barrier into the subfloor will prevent these issues from occurring. There's little reason not to add a vapor barrier, unless one is not available or you object on principle to the use of plastic.

is the current standard for most situations. There are also paint-on barriers available for use on concrete slabs. Often a vapor barrier is added in conjunction with insulation. If working with a pre-existing subfloor that may not include a vapor barrier, consider adding one or test to see whether one is present.

To test whether or not a concrete slab has a vapor barrier, tape a 2'x2' piece of plastic onto the floor for a few days. If there is no vapor barrier, moisture will build up on the underside of the plastic. This test is most telling in the wet season and in warm spaces. Pre-existing compacted gravel slabs are rare, as the technique is not widely known and is used mainly by people who are planning to pour earthen floors. Digging a small hole through the compacted gravel will reveal if the slab contains a vapor barrier or insulation.

To add a plastic vapor barrier, lay it down over the concrete or gravel, taking care not to poke holes in it or leave folds and wrinkles. A few wrinkles and folds won't be a problem if the earthen layer above will be 1½" thick or more; a layer this thick has enough weight to flatten out the folds and prevent the possibility of weak spots. If a thinner layer is desired, it's critical that the vapor barrier lie flat and smooth.

Bring the vapor barrier right to the edge of the room and fold it up the wall to just below the height of the finished floor. Tape it with plastic tape (duct tape works well) along any seams and to the walls. Tape around any pipes, heating ducts or posts that come up through the floor. Overlap seams about six inches.

Vapor barrier test on a concrete floor. (Credit: James Thomson)

Insulation

Many people mistakenly believe that thick, massive walls and floors provide good insulation. The reality is that a thick earthen wall or floor provides little insulation but lots of thermal mass. Insulative materials are light-weight and low in density, like a down sleeping bag. Mass is like a hot rock or hot water bottle you bring into the sleeping bag. The insulation keeps the heat that is slowly radiating from the rock or water bottle from seeping out too fast. The rock or water bottle is a battery, storing the heat from the fire.

Installed vapor barrier with seems taped, and taped to the wall.
(Credit: James Thomson)

Insulation is an important component of any floor system. Just as a camper sleeping outside needs a sleeping bag, a properly built earthen floor should be fully insulated to ensure an energy-efficient and comfortable building. The type and amount of insulation to be used will depend on the floor design and the specifics of the climate. Local building codes will specify required "R-values" for insulation in floors, walls and ceilings. "R-value" is a measure of thermal resistance; insulative materials have higher R-values. Insulation requirements for floors in the United States range from R-13 up to R-30 depending on the location. There are no code requirements (or even guidelines) for how much thermal mass to include. Consult with a building professional to determine the best choices for your specific situation.

On-grade applications (concrete or compacted gravel) require an insulation that can support the weight of a floor and the people living on it. The most common choice is high-density rigid foam specifically designed for use in subfloors. Only use a foam

Installing rigid foam insulation.

(Credit: James Thomson)

product that has the compressive strength to support the weight of floor loads! Typically this foam comes in 4'x8' sheets that have an R-value of 4 or 5 per inch. A spray-on product may also be available; consult with a local insulation installer. There are also thin roll-out foam products, often with a reflective side. These thin insulation products can be used as a vapor barrier and also add some R-value, approximately R-1 (or less) per inch. Do not put down multiple layers of this type of product; this could create too much bounce, which can result in floor damage.

Foam is highly insulative, but is a very energy-intensive product to produce. There are not many good "natural" options for insulating beneath on-grade floors. The two most commonly used options are pumice rock and perlite.

Pumice rock

Pumice is a naturally occurring volcanic rock that is lightweight and porous and can be found in a range of sizes and compressive strengths. The insulative properties of pumice will vary

Pumice can be used as an insulative base. (Credit: Ron Hays)

Pumice spread and compacted on the subfloor. (Credit: Ron Hays)

depending on its density; most types are approximately R-1.5/inch. Not all types of pumice are suitable for underfloor insulation. The recommended type has mostly larger (½" to 1" diameter) rocks and good compressive strength (it cannot be crushed easily). Smaller and softer types may be suitable but should be packed into bags so that the material cannot shift under the weight of the floor. If in doubt, do tests.

To install pumice, spread and level it over the vapor barrier, as thick as necessary to achieve the desired insulation level. Compact every 3 to 4" of material with a plate compactor or hand tamper.

Expanded perlite

Compacted bags of perlite used as insulation.

(Credit: Laura Bartels)

Perlite is a naturally occurring volcanic mineral similar in makeup to glass. When heated in a kiln, it softens and expands, almost like popcorn. Perlite has many uses in construction and agriculture. For insulation purposes it can have an R-value of up to R-3/inch. To install perlite, keep it in the plastic bags it comes in and place these

tightly together on top of the vapor barrier. Compact the bags with a hand tamper or plate compactor.

Important note: Regardless of the type of insulation, the finished layer of an earthen floor should never sit directly on insulation. There needs to be a dense structural layer of concrete, poured earthen basecoat or compacted gravel between the two, to create a firm, stable base for the finish layer.

Insulation between floor joists. (CREDIT: JAMES THOMSON)

Insulating under a framed-wood subfloor is a far simpler project, because the insulation does not need to be load-bearing. Simply install batt or "blown-in" insulation between the floor joists. There are far more options too for these types of insulation. Fiberglass is the most common, but more natural products made from wool, cellulose, recycled cotton and rockwool are also available.

Pouring over existing finishes

Sometimes there are other existing floor finishes installed on top of a wooden subfloor or a concrete pad. Examples would be tile (ceramic or vinyl), linoleum or hardwood. It is usually possible to pour over this existing surface, as long as the additional thickness of the earthen floor is not a problem.

Just like preparing for a new coat of paint, remove any loose material and scrape back until there are no loose edges. If the surface is unstable, consider laying down a separation layer, such as a sheet of plastic, or a reinforcing mesh to add tensile strength. As with any earthen floor installation, make sure there is no way for moisture to migrate into the floor from below. Install a vapor barrier if this is a concern.

Considerations for hydronic or electric floor heat

Note: Installation of hydronic tubes or electric mats must be completed after the subfloor is prepared and before pouring begins, and requires a few special considerations. See Chapter 13: Using Radiant Heat for details.

CHAPTER 7

Preparing the Work Area

ONCE THE SUBFLOOR IS READY, IT IS TIME TO PREPARE THE ROOM FOR THE FLOOR INSTALLATION, by protecting areas that could become damaged, setting up fans and dehumidifiers, creating level lines and screed boards and setting up the laser level (if available).

Marking the finished floor height

Start by lightly marking the finish height of the floor at several points around the perimeter. Use a chalk line or a pencil and a straightedge to extend the marks around the room.

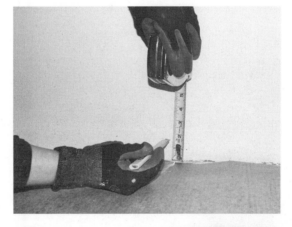

Marking finish floor height on the wall. (CREDIT: JAMES THOMSON)

Extending the floor height line around the room. (CREDIT: JAMES THOMSON)

Small gap at floor edge, a result of shrinking. (Credit: James Thomson)

Using baseboard and base shoe is a good way to cover the shrinkage gap at the edge of the room. (Credit: James Thomson)

Edge details

It is important to plan ahead for how the floor will meet the wall. Earthen floors shrink slightly as they dry, leaving a small space where the floor meets the wall. Conventionally framed structures are often designed to have baseboard trim with optional base shoe, which will cover this space nicely.

Some designs call for the floor to meet the wall without a baseboard, for example on walls that are curved or have heavy undulation. In this situation, steps can be taken during the burnishing process to minimize the space (see Chapter 9).

Preparing the threshold(s)

Determine how the earthen floor will meet the floors of any adjacent rooms. There are a few options for this, and much depends on the relative floor heights of each room and what is already in place or will be installed later. The easiest solution is to have floors in adjacent rooms at the same level as the finished earthen floor, so you can pour the earthen floor right against the edge of

Sukita:

A good trick for having the finish trim sit flush on a level earthen floor is to install a level board at the height of the finished floor before pouring. The baseboard trim will sit flat on this board. Make sure it is thinner than the thickness of the trim above it; otherwise it will show.

James:

For my bedroom, I installed the baseboard before I poured the finished floor. I did it this way because I wanted to put an earthen plaster in the room, and wanted it to land on top of the baseboard, so I needed to install the baseboard before the plaster. I wanted the floor to go in last, for simplicity and to prevent damage to it during the plastering and trimming process.

We attached the baseboard to the wall about ⅛" above where the finished floor would be. When we poured the floor, we could then tuck floor material under the baseboard. When everything was dried and oiled, I finished it off with a quarter-round base shoe, so the space between the bottom of the baseboard and the floor was not visible.

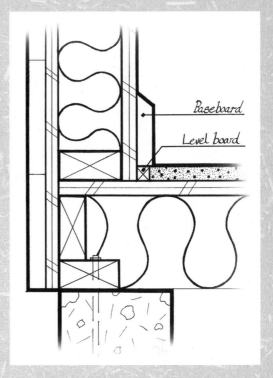

Illustration of level board at floor edge.
(CREDIT: JOHN HUTTON)

Base shoe used to cover vertical gap between baseboard and floor. (CREDIT: JAMES THOMSON)

the existing floor(s). It may work out that an adjacent room won't get its flooring installed until after the earthen floor is poured (especially if this is new construction). In this situation, install a stop to keep the earthen mix inside the room's boundaries (See diagram below left). This stop can be temporary (remove it once the floor has been oiled and is dry), or it can be hidden under a threshold sill.

An earthen floor meeting a tile floor (left) and a concrete floor (right), all at the same level, without any spanning thresholds.

(CREDIT: JAMES THOMSON)

Threshold profiles include:

- A wood stop as a temporary stop or nailer for a future attached threshold
- A spanning threshold that spans both flooring types
- A transition threshold that butts up against the earthen floor at one height and other floor type at another height

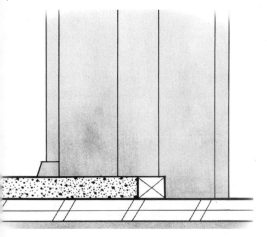

A temporary stop to pour fresh mix against.

(CREDIT: JOHN HUTTON)

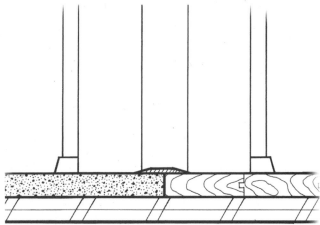

A wooden "spanning" threshold that covers the joint between the different floor types. (CREDIT: JOHN HUTTON)

- An embedded threshold, that creates a permanent joint between an earthen floor and another flooring material. A piece of angle iron in an L or preferably an I shape fastened to the subfloor works well for this.

It is also possible to slope an earthen floor up (or down) to meet a higher (or lower) threshold level. This slope should be gradual, typically spread out over several inches. A gradual slope is rarely noticeable in the finished floor.

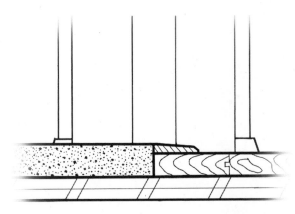

A wedge-shaped transition threshold joins two floors of unequal thickness. (Credit: John Hutton)

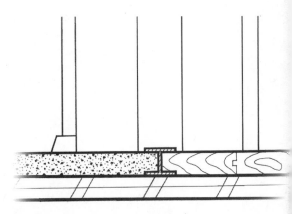

A piece of I-shaped metal can be used as a joint between flooring types. (Credit: John Hutton)

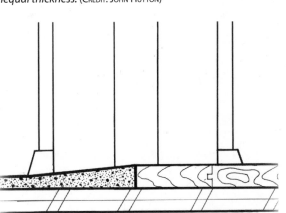

An earthen floor may be gently sloped at a threshold to account for differing floor thickenesses. (Credit: John Hutton)

Picture of an I-shaped metal strip joining an earthen floor to a cork floor. (Credit: Miri Stebivka)

These choices are mostly based on aesthetic preferences and practical requirements. However the threshold will be treated, prepare for it in advance by installing the threshold or a temporary stop if needed.

Protecting walls and trim

Masking plastic installed to protect finished wall during the floor installation.

(Credit: Sukita Reay Crimmel)

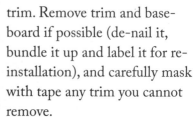

Earthen floor work is dirty! Wet mud and oil can stain walls and trim. Remove trim and baseboard if possible (de-nail it, bundle it up and label it for re-installation), and carefully mask with tape any trim you cannot remove.

Apply masking paper or plastic to the walls to protect them from mud and oil splatter. Plastic will protect against oil, though paper can be used if care is taken not to splatter oil (mud tends to splatter more than oil anyway).

If you are going to install a baseboard, apply the masking film from one inch above finish floor level (within the area that will be covered by the baseboard) to at least twelve inches up the wall. Use painter's tape to attach the masking film to the wall along the bottom and top edge of the masking film, and along any seams. The film must be well secured to the wall, because when the fans get turned on they can blow it right off.

If you are not going to install a baseboard, apply tape right at the finish floor level and plastic or paper masking above to protect the wall from splatter. To minimize the possibility of tape sticking to walls or peeling off paint, remove this tape after burnishing and reapply it before oiling.

Remove any HVAC registers in the floor area and cover duct openings. *(Note: Ducts and register covers may need to be extended to sit flush with new finish floor height.)*

Wood duct extension, built to match the finish floor height.

(CREDIT: MIRI STEBIVKA)

Preparing for proper drying

Pre-position fans, dehumidifier(s) and heater(s), if necessary. Consider pre-running extension cords and even building temporary platforms to hold fans up off the floor to facilitate proper drying. Direct fans to blow in a circular pattern around and out of the room. (See Chapter 8 for more discussion of this crucial step.)

Cover any windows that may allow direct sunlight to hit the floor while drying. Windows can be covered from inside or out. Direct sunlight

Painter's tape

A note here on tape. Painter's tape (often blue) is designed to stick to most surfaces and come off without leaving any residue or pulling off paint. But the longer the tape is left on the wall, the harder it will be to remove without damaging the wall finish. Because earthen floors take a long time to dry, the masking film could be on the wall for a couple of weeks or more. Inquire at a paint supply store about which kinds of tape work best for this situation, and be prepared to touch up the wall. The tape at the bottom of the masking just above the floor is likely to get wet, muddy and oily, which makes it even harder to remove without damaging the finish. Tape may need to be removed and replaced after each step if damaging the finish is a concern. Also, it is best to install (or reinstall) trim *after* the floor is finished to hide any damage caused by the tape.

Direct sunlight shining on a wet floor can cause irregular drying and thus cracking. Covering windows is recommended.

(Credit: Sukita Reay Crimmel)

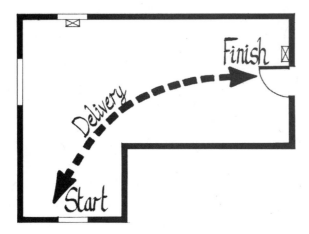

Delivery route. (Credit: John Hutton)

during the drying process can cause the floor to crack due to irregular drying (see Cracks section in Chapter 2).

Developing a pouring plan and material delivery route

Determine where the pour will start and end, and the route you will use to deliver the wet mud. A wheelbarrow works well to bring in the materials. Use cardboard to protect doorways and walls that may come into contact with the sides of the

*Protect door
jambs to prevent
damage from
wheelbarrows.*
(CREDIT:
JAMES THOMSON)

wheelbarrow. The route may also
need ramps at doorways and
protected pathways on the floors
through other rooms (use thick
paper or cardboard).

Preparing for leveling

Using a laser level is the easiest
way to ensure that the floor
stays level while pouring. It
is also possible to use a spirit (bubble) level with screed
boards.

If using a laser level, prepare reference sticks. Using a
wall-mount bracket or tripod,
place the laser level in a secure
spot so that it casts a line that

Ramps can make delivering material easier. (CREDIT: SUKITA REAY CRIMMEL)

Using a laser level to create a reference stick.

(Credit: James Thomson)

is approximately twelve inches above the finish floor level, visible to all the installers, and has few obstructions. (You may need to move the laser level periodically so that the laser line can reach all the work areas; pre-select locations to move it to if necessary.) Make a reference stick from a roughly 1" x ¼" x 18" piece of wood. Cut one end flat and hold with the flat end at the finish floor level mark on the wall. Use a small torpedo level to make sure the stick is plumb and mark where the laser

Sukita:

Not all laser levels are created equal. First of all, it is worth getting a high-quality, self-leveling one that projects a wide angle (360 degrees is the easiest). If the fan angle is less than 180, the level will likely have to be pivoted at some point to make sure the beam reaches the whole room. Use a wall mount and take time to ensure it is in a good location, so the beam will reach as much of the room as possible and is perfectly level. If the mount is even slightly off level, pivoting the laser on the mount will change the beam's height. When pivoting or moving the laser, make sure to double-check that the new line is at the same level as the old one. Take care of this sensitive tool!

Sukita's favorite laser level. (Credit: Miri Stebivka)

line hits the reference stick with a permanent marker. Be sure to hold the reference stick plumb when making this mark, and when using it during the pour! Extend the line so it is visible on all sides of the stick. Repeat this process to create a reference stick for each installer.

Preparing for screeding

Screeding is a common practice for pouring concrete. The same tools and techniques can be adapted for pouring earthen floors (see Resources appendix for recommended books).

Fresh mix ready for screeding.
(Credit: James Thomson)

Screeding is not always done, but it is especially useful if a laser level is not available, for hard-to-reach floor sections or in workshop situations where the installers are not as skilled with hand-troweling a flat floor.

Prepare for screeding by making rails the same thickness as the floor pour, by ripping boards to the proper thickness with a table saw. Screed rails should be about four feet long; make enough so each troweler has three or four to work with. Lay them out on top of the prepared subfloor.

Installing expansion joints, control joints and other crack-prevention measures

Some cracking in a finished earthen floor is normal. Cracks can be prevented, fixed or just left alone. Most cracks can be prevented by ensuring that the mix is of good quality (Chapter 5), and that the floor dries uniformly (Chapter 8). Large unbroken expanses of floor are more likely to crack from movement and shrinking than smaller floors. Expansion joints or control joints help to control the shrinkage and movement by breaking a larger floor into smaller sections, thus reducing potential cracks.

An expansion joint is a split or break that goes all the way through the layer and extends along the entire width of the floor. There are products available that can be installed to create this joint, or a thin piece of wood or metal can be cut to size onsite. Install these in advance of the pour.

An expansion joint creates a break all the way through the floor. (CREDIT: JOHN HUTTON)

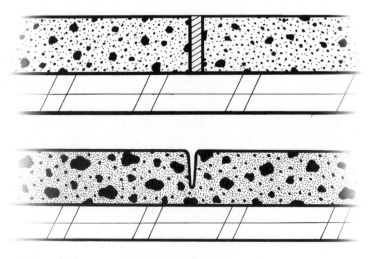

A control join goes part way through the floor. (CREDIT: JOHN HUTTON)

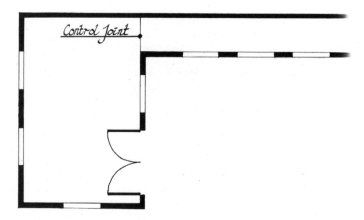

A control or expansion joint should be installed in locations where the floor narrows quickly. (CREDIT: JOHN HUTTON)

A control joint is a split that does not go all the way through the floor layer, but instead cuts through half or three-quarters of it. The joint is created by carving a groove into the floor while it is still wet. There are trowels designed just for this purpose. Consider installing an expansion or control joint when:

- installing on a wood-framed subfloor with a span greater than 15 feet;
- installing on a concrete pad that has an expansion or control joint in it;
- when a floor narrows quickly, at a hallway or threshold.

Areas that meet up against outside wall corners, posts or cabinets are also prone to cracking because these materials (usually wood) expand and contract at a different rate than the floor. To reduce the potential

A common crack resulting from a post in a floor.
(CREDIT: MIRI STEBIVKA)

Wrapping foam around a post base will help reduce the likelihood of cracks developing. (CREDIT: JAMES THOMSON)

for cracking at these locations, take the following precaution:

- Use tape to attach a ⅛"-thick piece of foam to the corner/post/cabinet from the top of the subfloor to just below the level of the finished floor. This creates a buffer against dissimilar movement, and possible cracking. Foam tape may also be used.

The mixing station

Chapter 5 outlines the process for determining a recipe and actually making the mix. Now it's time to do the mixing. To choose a mixing area, consider where the water source, materials and building entrance are located, as well as the mess created during the mixing process, the noise and

Look, no cracks! (CREDIT: MIRI STEBIVKA)

exhaust from the mortar-mixer and the effects of foot and wheel-barrow traffic. To aid in cleanup, consider laying out a large tarp under the mixing area.

For large floors with a big crew, the mixer will be almost constantly running to produce mix fast enough. Setting it up in an efficient way will save time and effort.

Sukita:

I walk around looking at floors all the time. I notice that the concrete slabs that are used as finish flooring often have an assortment of control joints and natural flowing cracks. The control joints do not always stop cracking. Personally, I love the look of the natural cracks. And when a crack appears in an earthen floor after the floor is oiled, the void can be easily filled, possibly with another color, to create a marbled look!

Cracks in an earthen floor. (Credit: Sukita Reay Crimmel)

SECTION 4

INSTALLING THE FLOOR

Finally after hours of gathering, processing and testing materials; preparing the subfloor with vapor barriers, insulation and radiant heat; and masking walls, protecting trim and pathways, it's time to start spreading mud on the floor! This is the part where the magic starts to happen, and the results of all the hard work thus far will payoff. After it's poured and dry, it will be sealed with oil, and maybe wax. Soon you'll have your own beautiful earthen floor to live on! Take it slow, and enjoy the process.

(PHOTOS: JAMES THOMSON)

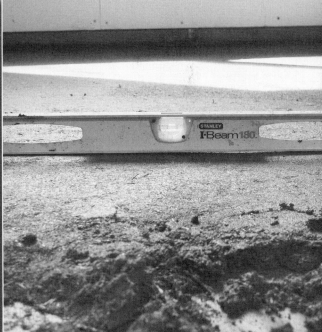

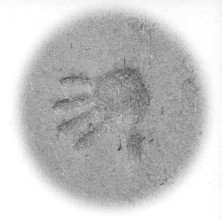

CHAPTER 8

Pouring the Floor

AT THIS POINT, THE SUBFLOOR SHOULD BE FULLY PREPARED as described in Chapters 6 and 7. The surface should be clean; earthen basecoats or compacted gravel subfloors should be free of loose pebbles or chunks of earth. Sweep and vacuum if necessary. For finish pours over earthen basecoats, lightly wet down the basecoat before pouring the finish layer.

The end result of this stage of the process will be a floor that is level, smooth and flat. Methods for achieving a level floor vary depending on the site and the tools available, but the basic application and troweling of the mix is the same for all of them.

Application and troweling

Deposit one to two buckets' worth of material onto the prepared subfloor, starting in the

Sukita:

It is not a good idea to mix or pour a floor when the materials are extremely cold. One cold winter day, I was working on a floor and had someone operating the mixer outside. The sun set about 5 pm that day, and there was one last batch to mix. As soon as the sun set, the temperature dropped and the consistency of the mix changed; it needed much more water to get to the same consistency as the previous batches. Even then, the final mix was much stiffer than the previous ones. Because of this, I stretched the mix a lot during the pour, which resulted in cracks later.

corner farthest from the room's exit (refer to pouring plan, Chapter 7).

Using a wooden float to spread and flatten an earthen floor.

Work the wet mix flat with a wood float to get it to the proper thickness. Be careful not to stretch the mix too far, "stretch marks" may result. Rework any area where this happens, as this is a place where cracking could occur during drying. Use a back-and-forth cutting motion to remove small amounts of mix from high areas without making holes in the floor, or add more mix to low spots and blend it in with the flat part of the float. Use the floor level guide marks on the wall to determine how thick to go. Work in small sections (about 2' x 2') and don't dump too much floor mix in one place at a time or there will be extra work moving the material that is in the way. Avoid kneeling in the pile of mix, which can create uneven compaction. Do not overwork the material; there will be an opportunity to get it smooth later. Check for level frequently.

Leveling

1) Using a laser level

If a laser level is available, you should have already set it up

"Stretch marks" in the wet mix, a result of stretching the mix too far.

and made reference sticks (Chapter 7). Rest the bottom of the stick gently on the surface of the freshly troweled area and hold it as vertical as possible. The laser line should line up with the reference mark. Adjust as necessary by adding more material or "cutting" it down, using the edge of the float with back-and-forth motions. Make sure that mud does not build up at the bottom end of the reference stick; this will throw off the measurements. For some floors, you will have to reposition or even move the laser level during pouring. As suggested in Chapter 7, use preselected locations to move the laser level to.

Using a reference stick with the laser level to determine floor height. (CREDIT: JAMES THOMSON)

Adjusting laser level so the beam reaches the other side of the room. Note the mounting bracket attached to the wall. (CREDIT: JAMES THOMSON)

Screeding wet mix. (Credit: James Thomson)

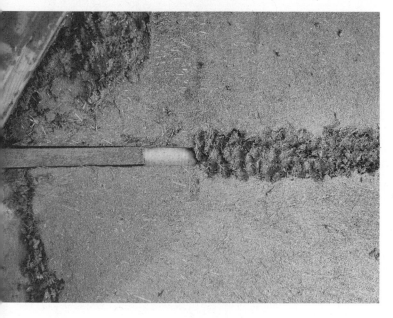

Working wet material into the voids created by the screed rails.
(Credit: James Thomson)

2) Using the screed board and rail technique

Place the screed rails before you deposit the mix. Use one at the perimeter, then another two or three spaced about every two feet. *(Note: The subfloor must be level for this technique to work!)* Spread the mix by hand to fill the space between the rails, then drag the screed board across the rails using a constant side-to-side motion. Fill any divots left behind, then repeat. Once the working area is smooth and flat, slide or reposition the screed rails into the next area to be worked on. Rough up the sides of the voids, fill with floor mix and use the float to blend, level and smooth the whole section.

Note: Leaving screed rails in the wet mix longer than twenty minutes can result in the areas around the rails drying out; when the new mud is mixed in, a cold joint and a possible crack can result.

3) Using a thickness gauge

Make a thickness gauge by marking a small stick (about the diameter of a chopstick and at least six inches long) with a marker at the same distance

Robert LaPorte's "wet screeding" technique:

My experience with earthen floors is very positive — no hesitation, I love them. It's important that your feet contact the Earth with nothing in between; it creates a very grounded feeling. Your body needs to be grounded to function properly. I built my first floor before I had ever seen one, in 1991. I really didn't have any information; I took what I had learned about natural plastering and essentially used that. It's a wonderfully forgiving technique. For beginners, start with a small section, figure it out there before you do the whole house. No worries, because you can do it in two or three layers if you want.

I like to do a technique I call "wet screeding." Say I have a twelve-square-foot floor. I'll start by pouring a perimeter along the edge, about a foot wide, using a laser level or spirit levels. Then I'll put a one-foot-wide strip down the middle of the room, again using the levels. The perimeter and the middle strip become the rails for the screed board. You have to be careful not to move the height of the original strips as you are screeding. I use a pretty dry mix, so it works well.

from the end as the thickness of the floor. Use the gauge to poke through finished sections of the floor while working, to ensure that the thickness is staying constant. When the end of the gauge hits the subfloor, the floor level should line up with the line on the gauge. Trowel over the holes when an area is checked and complete. This method only works if the subfloor is level.

Regardless of which method you use, check frequently with a four- or six-foot spirit level. Look for overall level and compare the flatness of the floor with the bottom of the level (which is perfectly straight). Some undulation is expected, but significant peaks or

Using a thickness gauge to measure floor height.

(Credit: James Thomson)

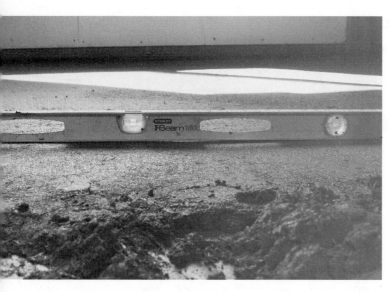

Using a 4' level to check for flatness while pouring. Note the gaps under the level; some irregularity is expected. (Credit: James Thomson)

valleys should be removed or filled. When working with a manual level, take care to keep muddy hands off the bubble glass or it will become impossible to read the level. Also clean off the bottom of the level often to ensure accurate readings.

Note: Do not spread out, flatten and level a larger section than you can reach across! Work in two- to three-foot strips. This allows you to go back and smooth out the material you have just laid down without having to walk on freshly laid material.

Working at a comfortable distance. (Credit: James Thomson)

Pouring at the edges and thresholds

If the floor is being sloped up (or down) to meet a threshold level (see Chapter 7), make this adjustment starting 6 to 12" from the threshold. For any edges where the floor will meet the wall without base trim, apply extra pressure while spreading and smoothing the material, to reduce the separation from the wall that may occur when the floor dries (Chapter 7). In some cases, the floor may need to be sloped up slightly at the edge to bring it to the proper height for base trim. This can be done as needed; gradual slopes will rarely be noticeable in the final product.

SMOOTHING

Note: This step can be skipped if this is a basecoat pour (see Chapter 6, sidebar).

Once a section is flat and level, swipe a steel trowel lightly over surface to make it smooth and glossy. Hold the trowel almost flat, but with enough of an angle that the trailing edge rests on the floor. Start strokes from the edge of the working area (the wall or the edge of the pour) whenever possible. If this is not possible, start moving the trowel over the surface and lightly settle it down while it is still moving (making a "soft landing"). Use broad arching strokes, working only as much

Avoiding cold joints

Cold joints can happen where already flattened and troweled mud meets new mud. It is important to keep the edges of the previously troweled area rough, and to blend the fresh material with the old material. If it's not blended well, a separation can occur at this junction during drying, leaving a crack.

Hold trowel at a slight angle and when possible begin the smoothing motion from the room edge. (CREDIT: JAMES THOMSON)

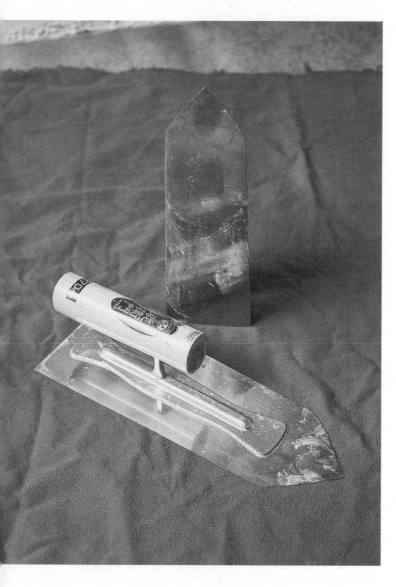

Delicate Japanese trowels are handy for smoothing tricky spots. (Credit: Miri Stebivka)

as necessary to smooth the surface and minimize trowel marks. Consider varying the size and starting/stopping points of each stroke so that troweling patterns do not develop (these patterns may be visible in the final product).

It should only take a couple of swipes with a steel trowel to create a nice, shiny surface. Don't overwork the floor — any trowel marks or other irregularities 1/16" deep or less can be taken out in the burnishing step. Small, more flexible trowels (see Chapter 4: Tools) are useful for tricky spots: corners, doorways or around objects in the floor (posts, fireplace hearths, built-in cabinetry).

Once a section has been smoothed with a steel trowel, it is done. Move to a new work area, dump some fresh floor mix on it and repeat the process from the beginning. Carefully blend the edges where the fresh material meets the already-smooth material, to avoid cold joints.

Once the next section is complete, there should be no visible seam between the two sections. Continue working in this manner until the pour is complete, usually at the threshold of the exit. There should be a threshold or stop in place already (Chapter 7); trowel right to this edge.

WORKFLOW

It's best to complete a pour in one day or cold joint cracks could result. Short breaks for lunch or snacks are usually fine, unless the room is especially hot and dry. If the floor is large (or the body is tired) and it needs to be completed over multiple days, or a longer midday break is desired and there's a risk of the working edge drying, be sure to prepare the working edge before the break so that it is ready for blending when work resumes. Smooth down the length of the edge to an angle with a steel trowel. This smooth slope keeps more moisture in and makes a "lap joint" with the next batch of fresh mud. Lay wet fabric or paper along the edge and at least eight inches back into the poured section and cover with plastic. When it's time to start working again, simply take off the covering, rough up the slope and work new mud into the edge. This area can be prone to cracking, so be attentive to work the new material well into the old.

Leave a sloped edge when taking a break for lunch or overnight. (CREDIT: JAMES THOMSON)

If there is mix left over (and there should be), save a gallon or two for burnishing, and dry another gallon or two (or more, depending on the size of the floor) for future repairs. Dry it out on cardboard or plywood in

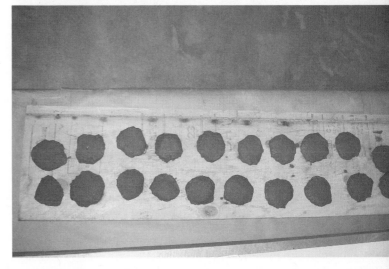

Drying floor biscuits, to use for future repairs. (CREDIT: JAMES THOMSON)

large cookie-sized patties. Collect the dry cookies in a bag and store in a dry place. Leftover mix can be disposed of in the yard, under a tree, even in the garden.

Seal off the room so that people and especially animals cannot gain access. Make sure open windows have screens, and that the lower halves of doorways are firmly blocked with plywood.

Turn on the fans and allow the floor to dry for twelve to forty-eight hours, just long enough so that it sets a bit but does not dry through.

Michael G. Smith:

It was the final day of a floors and plasters workshop. We poured the finished layer with a beautiful red clay mix and left it overnight to dry. The next morning we went to look at it, and there was an incredible mess on the floor. The first thing I saw was cat footprints all over the place. Then, in the middle of the floor, a pile of mouse guts. On closer inspection we could see mouse footprints, you could track the whole chase! Fortunately we were able to get back out on the floor and burnish out the prints.

I've had something similar happen so many times. Whenever you put in a floor, act as if every cat within ten miles is headed there right now. I've had cats who were like building inspectors; the day I built something they'd have to come and walk around on it. I've also had little kids jump right in a fresh pour. All kinds of feet end up in wet floors. Sometimes it's nice to leave the prints in, as a memento....

A cat has left her mark in a fresh floor.

(Credit: Sukita Reay Crimmel)

Burnishing

Burnish (verb) --
1. To make shiny or lustrous especially by rubbing. Polish.
2. To rub (a material) with a tool for compacting or smoothing.

Burnishing the finished floor with a hard steel trowel gives it a polished look and compacts the material to make a harder surface. It is not a required step, but is recommended for increased longevity and attractiveness of the floor. It is best to burnish while the floor is still slightly wet.

The burnishing process

Materials and tools required: Steel trowel, floor pads (or pans), mister, extra floor mix

Burnishing with a steel trowel, using stainless floor pans.

(CREDIT: SUKITA REAY CRIMMEL)

Test an easily accessible area of the floor to see if it is dry enough to work on. It should be able to withstand firm pressure from a trowel without moving much or pulling apart. Try walking out onto a foam pad or steel tray. If it's possible to make a dent deeper than about ⅛", the floor is still too wet and will need more time to dry.

Once it's ready, carefully walk out to the far end of floor, using foam pads or steel floor pans. Mist down a 2'x 2' section of the floor with the spray bottle (place this on a pad or pan when not in use to prevent it from denting the floor).

Working on the wet floor on foam pads. (Credit: James Thomson)

Using a second trowel for balance to reach further out on the floor.
(Credit: James Thomson)

Smooth out the floor by applying pressure with a steel trowel, reducing the high points and compressing the floor. This requires a substantial amount of force! The surface should be shiny and smooth when done. If there are areas that are particularly low, add a little wet mix to fill them out.

Continue working toward the door. The floor will have wetter and drier parts, requiring different amounts of water from the spray bottle and pressure from the troweler.

There's a lot of delicate balancing that goes on during the burnishing stage, because care must be taken not to walk or place objects on the floor. Use the foam pads like movable stepping stones to reach the various sections, and a second trowel for balance while reaching across the floor to burnish areas farther away from the pads. This is the opportunity to make the floor perfect!

Burnishing the edges

Pay extra attention to how the newly poured earthen floor meets the wall. As mentioned in Chapter 7, the floor will

Sukita:

A student from one of my earthen floor workshops recently showed me a photo of her using foam pads in a unique way. She duct-taped smaller pieces of foam to the bottom of her shoes to get out onto the wet floor. This only works to travel across the floor — any kneeling down to do work on the floor would require removing the tape — but the foam "shoes" could be useful to reach that forgotten light switch or to turn on a fan.

Anne shows off her foam pad shoes.

(Credit: Kata Palano)

shrink and pull away from the walls as it dries, leaving a space. If baseboard trim will be installed, then no special treatment is needed at the edge. If no trim will be used, apply extra pressure at the edge to push material back into the space and firmly pack it against the wall. A separation may still develop when the floor dries; this can be treated like a crack and filled (see instructions at the end of this chapter). Tape should not be left on the wall for more than two or three days anywhere it will be visible as it could damage the wall

Sukita:

I have not always burnished floors. When someone first suggested it, all I knew of burnishing was with a small stone. The idea of burnishing a floor this way was tiring to even think about. Using a steel trowel was much better, though still labor intensive. I have burnished many floors after they are dry, and the pattern of the trowel has always shown up, so I now keep that in mind and created patterns intentionally. Burnishing when the floor is still a bit damp is my preference. Trowel marks and any dings from the pour day are easier to remove, and the end result is a harder floor.

finish, so remove the tape after burnishing and reapply tape and masking film before oiling.

Note: If the floor is dry to the point where it cannot be smoothed out effectively, wet the surface down with water from the spray bottle. This is an indicator that the floor dried too much before burnishing. It's a little harder to burnish it at this point, but it can still be done with good results, though the surface will have more of a marbled look with less compaction. Just keep misting and burnishing until it looks right.

Once the burnishing is complete, the floor can be allowed to dry fully. Seal off the room to protect it from curious humans and other animals.

Drying the floor

A freshly poured earthen floor contains a lot of water. Drying the floor properly and quickly is critical for reducing the possibility of mold growth or cracking. The amount of time it takes to dry thoroughly will depend on several factors, such as how thick the floor is, how wet the mix is, the indoor and outdoor temperatures, humidity and airflow.

Ideally the floor will dry uniformly. When clay dries, it shrinks. If one spot of the floor is drying while an adjacent spot is still wet, the two will have different rates of shrinkage, which could result in cracking. Make sure any windows that might allow direct sunlight to shine onto the drying floor are covered. Drying occurs as water evaporates and is carried away from the floor (via air). If it is dry and warm outside the house, simply leaving the windows and doors open and a fan or two moving air around will be enough to ensure good drying. In some climates, the air is so dry that no fans are needed, and the doors and windows should be kept closed so as not to dry the floor too fast.

If the outdoor temperature and/or humidity level aren't cooperating (which is often the case, as floors are usually the last step of the building process, which means they are being installed in the late fall after a long summer of building), mechanical assistance from fans, heaters and dehumidifiers will be necessary.

Sukita:

It is common to see white fuzzy molds growing on the little straw pieces at the surface of the floor during the drying phase. Do not be concerned. Once the floor is dry, they will stop growing, and you can simply gently sweep away any remaining bits. And in my experience the oil hides any discoloration caused by mold growth. Mold spores are everywhere, and most molds are harmless to most people. Earthen floors introduce a lot of moisture into a space, so it is important to take steps to dry the floor in about a week. If it takes longer and mold starts growing, it's not that big a deal, because once the moisture is gone, mold growth stops. If you have any concerns, consult with a mold mitigation professional.

Temperature: Keep an indoor temperature of at least 60 or 65 degrees F. If it's cold and the home has central heating, just set the thermostat appropriately. If not, space heaters may be required. If radiant floor heat has been installed, set the water temperature to 90 degrees while the floor is drying. This is likely lower than the normal working temperature of a radiant heat system (usually about 120 degrees); once the floor is dry, it can be increased to match the design temperature for the system. The oil and wax will cure faster when there is more heat. See Chapter 13 for more on radiant heat.

Airflow: Increasing airflow over the wet floor with fans is by far the best way to speed the drying process. Place fans strategically before pouring, as it will not be possible to walk on the wet floor to position them or turn them on! Use extension cords, switched outlets or even the home's circuit breaker panel to keep the hard-to-reach fans off while working.

Aim fans to blow air over the floor and out of the room through a window or door. Fans will need to be placed where they won't sit on the wet floor. Use existing features such as windowsills or built-in shelves, or hang the fans with string or build a temporary shelf somewhere in the room.

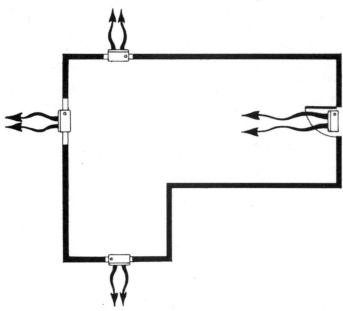

An example of fan placement. (Credit: John Hutton)

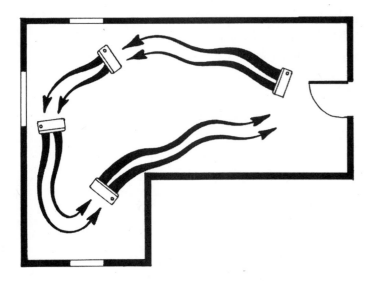

Creating an air "loop" on partially dried floor. (Credit: John Hutton)

Once parts of the floor have dried, the fans can be placed on the dry areas to reach the parts that are still wet. Positioning them in a looping pattern can speed the drying process. Just be careful — the floor is very fragile before it is oiled and sealed.

Dehumidifiers

Dehumidifiers are useful if the humidity level outside is high (Portland, OR, in the winter, for example), or there is limited airflow (a room with few windows or doors). Types of dehumidifiers are discussed in Chapter 4. Think about where the dehumidifier should be positioned before starting the pour. The ideal location is as close to the middle of the room as possible, but more practically it will need to be at or near the entrance.

Drying time

As mentioned, the total drying time will vary, and could be anywhere from a couple of days to a couple of weeks. It is best to dry the floor within a week, so any grain seeds that did not get removed during the straw sifting don't sprout (see below, "cracks and sprouts") and mold does not have a chance to get established (see sidebar page 151). It must be thoroughly dry before applying oil and wax.

Cracks and sprouts

Sometimes, in spite of the best efforts to prevent cracking or sprouting, they happen anyway. It's not uncommon to come back after the floor has completely dried and find a few small cracks or a sprout or two that have pushed up through the surface and created a small divot in the smooth finish. The best time to fix these issues is after the floor has dried but before the surface gets oiled and sealed. A few techniques for doing this are described below.

When performing these repairs, take care not to overwork the floor with the trowel. Only do the minimum amount of smoothing and pressing needed to seal up the crack or divot. Burnishing after the floor has dried will leave a different appearance on the surface, which could make the repaired area stand out from the rest of the floor.

For hairline cracks and blemishes: Use a spray bottle to mist down the area. Then use a steel trowel to press down the area, pushing the crack back together or smoothing over the divot.

For medium-sized cracks and divots (hairline to ¼"): Sift some dry floor mix through a fine window screen. Then lightly mist the area with water and brush the dry mix into the crack or divot, filling it up. If crack is deeper than ¼", layer light mistings of water with the dry floor mix until the void is filled. Mist and press the material down, without overworking it.

For large cracks and divots (¼" or larger): Fill large cracks with some wet mix, taking care to wet the sides, rough up the edges and work the filler mud into the sides. It may be easier if the mix used for the repairs does not contain straw. Allow to dry somewhat, then lightly burnish and blend as necessary.

For sprouts: Small sprouts from seeds will push up through the floor and disrupt the smooth finished surface. The sprout will die once the floor dries, but it's best to take them out and smooth over the area they pushed through. The damaged area can be treated like a small divot. This can be time-consuming work if there are a lot of sprouts, another good reason to sift the straw before using it (Chapter 3).

On rare occasions, the floor cracks and sprouts in enough places that it's worth treating the surface of the whole floor with a color wash, to cover the irregularities that result from fixing the cracks. See Chapter 11 for more on color washes.

An extreme example of sprouting in a slow-drying floor. (Credit: Sukita Reay Crimmel)

Sukita:

I have had a few floors that have cracked and some that have sprouted! After the work of fixing these mistakes, what remained were floors with some irregular areas, producing a visually unsightly, though structurally sound, floor. Applying a color wash the same color as the actual floor has been my favorite trick in these situations. This technique reduces the amount of straw that is visible in the finished floor. Sometimes I will apply the color wash with a brush only, and sometimes I follow the brush with a trowel. The pattern of the brush or the trowel will show. I am always careful not to over-work the surface, which can result in burn marks from the trowel or small cracks in the clay wash.

Brushing a color wash onto an unsealed floor.
(Credit: Miri Stebivka)

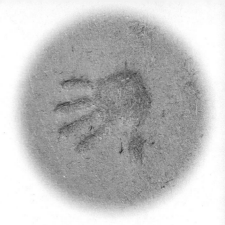

CHAPTER 9

Finishing the Floor

THE FLOOR LOOKS BEAUTIFUL, but in its unsealed state, it is very fragile. The finishing steps are what transform the unstabilized earth into a durable and waterproof surface. Without oil, the floor would quickly become damaged and dusty if it was walked on. After sealing, it will survive many years of use.

Oiling the floor

Oiling the floor with linseed oil is the step that gives an earthen floor its hardness and resistance to moisture. The oil is absorbed by the dry earth mix and polymerizes as it dries, forming a sturdy barrier and holding the material together. Without the oil, an earthen floor would simply be dry mud. The oil turns it into a durable and cleanable surface that will last for many years.

It is very important to understand the safety precautions required when working with oil. Read the section in Chapter 3 on linseed oil and the Oil and Wax Safety appendix. If you still have questions about performing this step safely, consult with a professional.

Preparing for oiling

Once the floor has dried, it can be walked on carefully. It is best to proceed with bare feet and rolled-up pants to prevent scuffing.

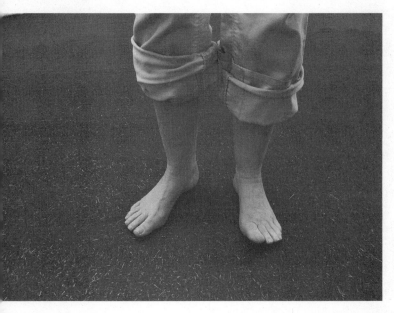

Sukita demonstrates proper cuff rolling for working on an unsealed floor.

(Credit: James Thomson)

Before applying the oil, protect the walls from "oil splatter," as oil will stain wood, stone, plaster and concrete. If the wall protection applied prior to pouring is still up, check to see that it is still well attached and covering the areas it needs to. If it's been removed or has fallen down, put up more. Remove any large bits of debris that may have landed on the floor, but don't sweep it. The oil will not bind down any small loose grains of sand or clay, so these can be swept up after the oil has dried. Sweeping prior to oiling can scratch the floor.

You should have prepared or purchased the oil already (Chapter 3). One gallon of oil will cover 35–45 square feet, usually in four coats. Make sure there's enough oil to complete the job, as all the coats are applied on the same day (the oil does not need to dry between coats). It can be applied with either a brush, a paint roller, rags or a sprayer. Gather the appropriate tools for the floor size and resources available.

Note: A floor wet with oil can be easily dented as it is curing. When working on the freshly oiled floor, keep pant cuffs rolled up so as not to scratch it (or stain clothes!). Set the tools down gently, being careful not to make scratches or dents. Use a rigid foam pad as a staging area for the oil bucket and tools. Keep plenty of clean rags and some solvent handy to clean hands, feet and tools when finished. Oil can stain anything it comes in contact with!

Applying the oil

Begin by pouring off some of the oil into a smaller container (one-quart yogurt containers work well) for easier handling. Use a brush to apply oil to the edges of the floor, covering a strip four to six inches from the wall.

Note: When oiling a floor edge that connects directly to the wall with no baseboard trim, hold the brush back from edge ⅛"–¼" and allow the oil to move on its own to the edge (this is especially important if there is a porous wall finish, like earthen or lime plaster).

The first coat can be applied fairly thickly; the floor should absorb most of the oil. Once the whole perimeter has received one coat of oil, apply another coat immediately. Repeat until three coats have been applied. If there are two people working, one person can continue around the edge while the other starts to oil the middle of the floor, called the "field."

Apply oil to the field of the floor with a brush, roller or rags. A paint roller with an extension pole works well for large floors and allows the installer to remain standing while working. Use a bucket with a roller grill or a paint pan for added efficiency when using a roller. Rags are a fine low-cost solution, but take care not to rub the floor, as

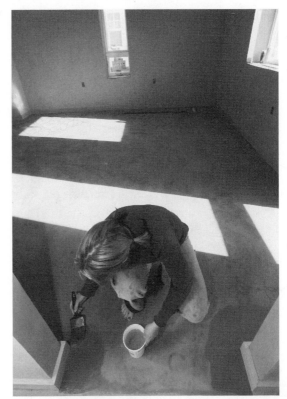

Oil the edges with a brush.
(CREDIT: ALAN S WEINER)

Applying oil to the main part of the floor with a roller.
(CREDIT: ALAN S WEINER)

it is easy to damage the surface at this stage. For very large floors, or for contractors looking to reduce installation time, an oil sprayer can be used (see Chapter 4). When using a sprayer, prestrain the oil through a fine screen so that any small particulates do not clog up the spray nozzle. Clogging will reduce the misting capability of the sprayer, and you will get a single stream of oil, which will slow the process.

Just like with the perimeter, three coats can be applied to the field immediately one after another. This will require walking on freshly oiled sections of the floor; this is not a problem as long as you do not walk on the final coat. Each coat will require progressively less oil to cover the floor. The oil pattern is easy to see during the first coat. Pay close attention to where you have oiled with the next few coats.

Darker areas have more oil than lighter areas. A brush tends to apply more oil than a roller or rags, so the perimeter may have more oil than the center of the floor. After applying three coats to each part of the floor, look to see if the perimeter is darker than the field; if it is, apply another coat to the field. The whole floor may show darker and lighter areas, and possibly even spots where oil has dripped off a bucket or brush. These irregularities will typically balance out and not be noticeable once the oil has dried.

Apply oil until the floor is saturated, usually about 4 coats. The saturation point will become apparent when the floor stops absorbing oil, and oil starts puddling on the surface, usually in an irregular pattern.

Oil tends to penetrate in an irregular fashion, as seen here.
(Credit: Sukita Reay Crimmel)

Stop applying oil to the areas where this occurs, and mop up pud-
dled oil with a rag.

If using a sprayer: Make sure the wall is well protected, as the
sprayer is harder to control at the edges. If the masking is done
well, the edges won't need to be brushed separately. Use the sprayer
everywhere on the floor and apply one coat right after the other.

Cleanup: Dry and dispose of oily rags as described in the Oil and
Wax Safety appendix. Brushes can be cleaned with a solvent (tur-
pentine, mineral spirits or orange oil); rollers should be put outside
to dry and then thrown away. Pour leftover oil back into its original
container for storage. As long as it is well sealed, oil will last for
many years and can be used for touch-ups or new floors in the future. It is common for some sand transported from the brushes and
rags to build up at the bottom of the oil; this is not an issue.

If there is masking tape on sections of wall or trim that will not
be covered by baseboard, remove it now. It is difficult to remove oily
tape once the oil has dried! Reapply tape and masking film before
waxing, as wax can splatter when it is buffed.

While curing, the solvent that is mixed with the oil evaporates
and releases VOCs, so the room or rooms should not be used for at
least two days. Seal these rooms off from the rest of the building,
open doors and windows to the outside, set fans on the floor and
blow air and fumes out. The VOCs will evaporate after two to three
days with good ventilation. See Oil and Wax Safety appendix for
more discussion of VOCs.

Temperature: Keep the indoor temperature to at least 65 degrees
F. If it's cold and there is central heating, just set the thermostat
appropriately. If not, space heaters may be required.

Allow the oil to cure for at least seven days before proceeding
to waxing (or moving in if wax will not be used). Make sure there
is adequate ventilation during curing; leave windows open and use
fans to circulate air with the outside. It is no longer necessary to use
a dehumidifier, but heat in the room helps.

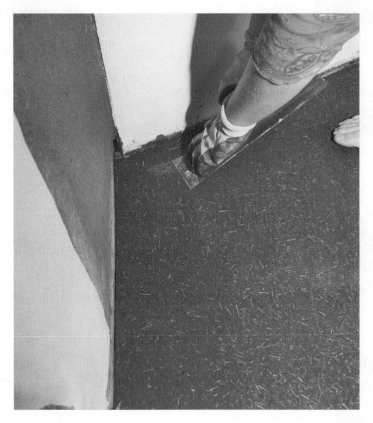

Using a stiff trowel to check the hardness of a freshly oiled floor.

(Credit: James Thomson)

James:

We hope we do not confuse readers by including steps that are optional and raising the inevitable question: "If it's optional, why do I need to do it?" These optional steps primarily change the aesthetics of the floor, and add extra protection that some people might find comforting. It's like getting the leather seats or the Scotchgard fabric treatment for your new car. Not necessary, but depending on your tastes and habits, maybe desirable!

Testing that the oil is cured

Walk barefoot (with pants rolled up) to an out-of-the-way spot with a stiff steel trowel and a foam pad. Kneel down and firmly rub a small section of the floor with the trowel. If there is any sort of movement, denting or pulling apart, the floor is not ready. If the floor stays solid, it is cured enough for use, or for further finishing (see below). The oils will continue curing for several weeks, but the floor can now be lived on. Many people choose to stop here, and enjoy many years of use. There are additional optional steps described below that will give a smoother, glossier finish and add an extra layer of protection (wax), but are not necessary for a usable and beautiful floor.

There may now be some cracks visible on the floor that weren't present (or noticed) before oiling. These can be fixed now, using the process outlined in Chapter 10.

Optional steps: Sanding and waxing

Sanding

Once the oil is cured, the floor can be sanded for a smoother

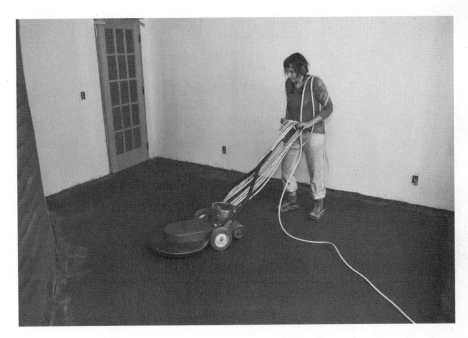

*Sanding the floor
with a high-speed
buffer and green pad.*
(CREDIT: JAMES THOMSON)

finish. This step removes some of the rougher texture and leaves a smoother, more polished finish.

There are several tools that can be used for sanding (described in Chapter 3: Tools). For small floors, use an electric orbital sander or angle grinder. For larger floors, consider renting a floor buffer machine or a large square orbital floor sander. Use a green scrubby pad (preferred) or 80- to 120-grit sandpaper. Lightly sand the whole floor, starting around the perimeter with the hand sander and switching to a larger tool (if available) for the middle. A light pass over the floor is all that is needed. The sanding process may have a negligible visible effect on the floor (especially if using a green pad instead of sandpaper), but by feeling it, it should be easy to tell which parts of the floor have been sanded and which have not. Parts of the floor may

Sukita:

I started sanding the oiled floor a few years ago, and it took my floors to another level. After just a single pass of a sander or buffer, the surface feels significantly softer to the touch, as it might feel after years of good foot polishing or lots of dancing.

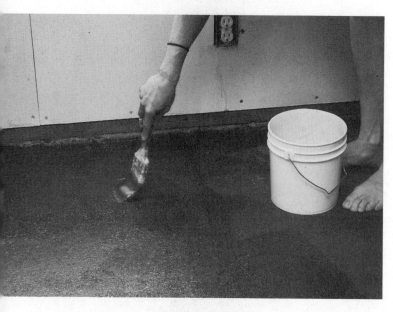

Applying wax to the edge of the floor with a brush. (CREDIT: JAMES THOMSON)

Using a sheep's wool applicator with an extension pole to wax the main part of the floor. (CREDIT: JAMES THOMSON)

scratch a bit during this process, but the waxing step will remove any scratches, so do not sand the floor unless you intend to apply wax as well.

Note: The larger buffing and sanding machines are tricky to operate, and should only be used by people who are familiar with doing so. Get professional support if needed.

Waxing and buffing — wax on, wax off!

APPLYING WAX

Wax may be applied whether or not the floor is sanded as described above. Waxing adds a glossier finish to the floor and additional protection against moisture. Floors in wet areas like bathrooms or kitchens benefit from a coat of wax.

The recommended floor waxes contain some linseed oil and solvent, so use the same safety precautions outlined in Oil and Wax Safety appendix. One gallon of wax will cover 340–350 square feet of floor.

Gather the tools: A brush, a sheep's wool applicator (or soft rags, or a sponge), and some rags for cleanup. Make sure the walls are still protected and that the floor is free from debris. Reattach wall protection, and sweep and vacuum if necessary.

Pour the wax into a dish tub-sized container and place it on a foam pad or towel to prevent drips from getting on the floor. Use a brush to coat the edges of the room and the sheep's wool applicator for the rest of the floor. Apply the wax thinly over the entire surface. Work from the back of the room toward the exit, taking care not to splash wax on the walls or walk on the freshly applied wax. To help ensure an even wax finish, keep up or put back up window coverings to block the sunlight from drying out one area more than another. Ventilation is also a priority; make sure that any sunshades do not block the flow of fresh air.

Clean brushes with solvent and dispose of the applicator and rags in the same manner as the oil-soaked rags.

BUFFING THE WAX

The purpose of buffing is to spread the wax deeper into the floor and remove excess wax to create an even finish. Allow the wax to cure before final buffing; heat and ventilation will help it cure faster. While the wax is curing, the room(s) should not be used or lived in — seal them off from the rest of the house, open doors and windows and set fans in windows or doors to pull the air and fumes out. The VOCs are usually gone after two days with good ventilation (see Oil and Wax Safety appendix for more discussion of VOCs). If the air is dry and

Buffing the floor with a hand-held grinder and custom-made white buffing pad. (Credit: James Thomson)

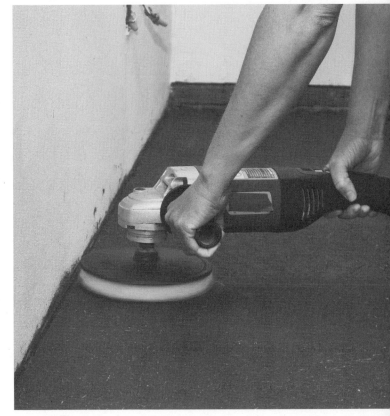

warm, the wax will cure quickly, within two days; cooler weather will require a longer drying time. Feel the floor daily to check its progress. If it is starting to feel dry and tacky, and appears dull rather than shiny, it is ready to be buffed.

Buff the wax using a large floor buffer, square floor sander or angle grinder with a white buffing pad. Some parts of the floor will be stickier or waxier than others; use the buffer to spread the wax and ensure the floor is evenly coated. If the floor is still slippery after the first pass with the buffer, buff it again to remove any excess material.

After buffing, allow the wax to dry with fans going for a few more days, up to a week if possible. Once the wax is dry, the floor will feel dry and be ready for use. It will continue to cure and harden over the next several weeks and months, so take care in the beginning not to dent it with heavy furniture or other point loads until it has reached full hardness.

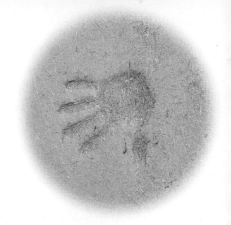

CHAPTER 10

Living on, Maintaining and Removing an Earthen Floor

A WELL-MAINTAINED EARTHEN FLOOR will last for many, many years. Like many materials, it will wear and change over time and require periodic minor repairs, and occasional major maintenance. The best way to keep a floor looking like new is to clean it regularly and limit the risk of denting and scratching. Heavy point loads like furniture, bits of rock and gravel stuck in shoes or even stiletto heels can cause little dents. Use floor protector pads on the feet of all furniture and encourage guests to leave their shoes by the door. This makes for less cleaning, plus it feels good to walk on earthen floors without shoes.

For cleaning purposes, earthen floors can be treated like

This type of furniture pad is used with wheels. Furniture without wheels more commonly has a pad stuck to the bottom of the leg.

wood floors. Use a broom or vacuum for regular cleaning; mop with a damp mop and an oil-based soap when a deeper clean is needed (see Resources appendix for product recommendations). Just like with a wood floor, it's not wise to leave standing water on an earthen floor for very long, so be judicious with the mop water!

Refinishing

Depending on how heavily it is used, a floor may need periodic refinishing with oil or wax. Refinish with the same oil or wax originally used. Heavily used floors (mopped daily, for example) may need to be refinished every year or two; most floors will only need it every three to six years. If the surface starts to look dull and dry, or if water does not bead up on it, apply a new coat of finishing treatment.

A floor can be refinished with a fresh coat of oil, wax, or both. Aggressively scrub the floor with soap and water and let dry. For floors in particularly poor shape, sanding, as described in Chapter 9, is also an option. Apply the wax or oil as described in Chapter 9. It is not advised to live in or use the space for a few days after applying oil or wax, to allow time for the VOCs to evaporate.

Maintenance: Cracks, dents and chips do occur occasionally. All of these can be fixed, though usually the patch will always be visible.

Dents: A "dent" is a divot or depression that does not break through the finished (oiled) surface. These can occur under point loads (a chair leg, for example) or where something heavy is dropped on the floor.

Dents in a floor.

(Credit: Miri Stebivka)

There is no need to repair a dent, as there's no risk of it getting worse, and its appearance is probably less noticeable than a repair would be. Dents add character!

Chips: A "chip" describes an area where the finished oiled surface has broken off, sometimes all the way down to the raw earth (½" deep). It is important to repair a chip (see Repairing chips and cracks, page 170), because the chipped area represents a weak spot on the floor. Left unrepaired, this damaged area will get bigger.

If you cannot repair the chip right away, apply finish oil to the area so that it does not take in moisture. You can repair it fully at a later date.

Cracks: As discussed, cracks happen either shortly after installation or years down the road. Small hairline cracks can be filled with extra wax, or just left alone. Wider cracks do pose a risk in that they often expose some of the unsealed material below. With time and continued foot traffic, a crack of this nature could expand.

Chips in a floor. (CREDIT: MIRI STEBIVKA)

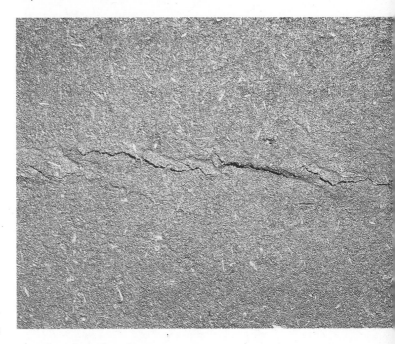

A small crack in a floor. (CREDIT: MIRI STEBIVKA)

A crack after a repair.

(Credit: Miri Stebivka)

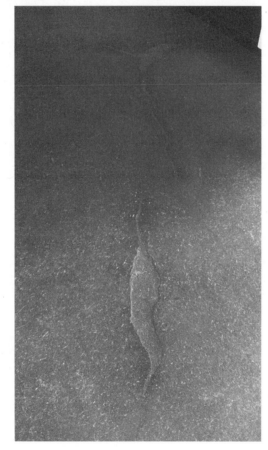

This crack was shaped before being filled to provide an aesthetic accent.

(Credit:

Jahmez Reismiller)

Repairing chips and cracks

It's easy to patch chips and cracks, but almost impossible to blend a patch into the surrounding earthen floor perfectly so that it is not visible. A small crack, properly filled and sealed, will be very hard to see, but will still be visible to the discerning eye. Some choose to add pigment to the patching material, or shape the crack in some pleasing way, so the crack becomes an aesthetic feature. If there are several chipped or cracked areas near each other, consider removing and refinishing a larger section of floor.

Take precautions to protect trim, walls and other items that might get dirty or damaged during the repair process.

Patches are made from the dried mix saved from the initial installation. (If none was saved, more will need to be mixed. Hopefully the original recipe is still lying around...!). Pulverize the dried mix into a powder and sift it through a fine window screen (this makes filling small holes and cracks easier; the step can be omitted if the area to be patched is larger than one inch square). The dried mix will then be mixed either with water, to make a material much like what was originally used to pour the floor, or with the finishing oil, to make a paste.

Clean the surface area around the chip or crack with a green scrubby pad or 80-grit sandpaper, up to at least four inches around it.

For large chips: Remove the loose material in the chipped area, and clean up the edges of the chip so that they are smooth and angled back into the floor, and the bottom of the hole is wider than the top. Mix the dry patch material with water, and wet the inside surfaces of the chipped area with a sponge or sprayer. Then smear some of the patching mix into the hole and level it with the surrounding floor using a wood float. Make sure the patch fills the entire void out to the edges of the damaged area. Use a steel trowel to smooth the surface of the patch; any scratches will be taken away with the oil. Wipe away any excess material on the undamaged floor with a rag.

Make sure to section off that area of the floor so no one disturbs the patch while it is drying. Once it has set up a bit (6–24 hours), burnish it with a metal trowel or the lid of a

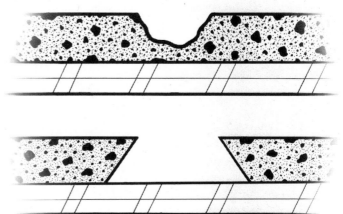

Top: cross section of a chip before repair; bottom: the edges of the chip are shaped at an angle to best hold in the repair. (CREDIT: JOHN HUTTON)

The lid of a yogurt container with the edges cut off makes a great burnishing tool for small patches and hard-to-reach areas. (CREDIT: JAMES THOMSON)

The materials (and tools) needed to make a crack-filling paste of oil and earth. (Credit: James Thomson)

Using a steel trowel to smear the oil and earth paste into a small crack.

yogurt container. Allow it to dry fully, then brush on up to four coats of the finishing oil. Cover some of the undamaged area around the patch with the fresh oil too. Then allow to dry another five to seven days. Once the oil has set up, it can be sanded, buffed and waxed just like in the installation process (Chapter 9). Refinish the floor around the patched area too to make sure the transition is well sealed.

For small chips (less than a couple square inches) and cracks: Smaller holes can be filled with a paste of dry mix and oil. The advantage to this technique is that you don't have to wait for the patch to dry before oiling (as above). Follow the guidelines in the Oil and Wax Safety appendix, and be sure to clean all tools of the oily mud mix.

Clean up the damaged area and the surrounding floor as described above. Then coat the inside surfaces of the damaged area with oil. Use a steel trowel to carefully press the patch material into the void. Smooth the surface with the trowel, wipe up excess material and let dry for seven to twelve days, until the surface is hard enough to take heavy pressure from a steel trowel and not move. Finish the patch with sanding, waxing and buffing (as described in the installation instructions) if desired.

Recycling and disposal

Excess floor mix can be used again on future floor projects. It can also be used for other natural building applications, like cob walls or earthen plasters; just note that it contains less clay than cob or plaster typically need, so add some extra clay. For long-term storage of floor mix, allow it to dry fully (best if done in a thin layer so that it dries quickly), break it up into pieces and store in a dry place. To dispose of it, just spread it in the yard; it will eventually blend in with the ground. Floor mix also works great to fill holes in the yard. Large amounts of material can be deposited in a "landfill": a place that accepts soil and dirt from excavation sites.

Removing an earthen floor. The top oiled layer will come out in chunks.
(CREDIT: SUKITA REAY CRIMMEL)

Floor removal

Sometimes a floor has served its useful life, or design changes for a room mean that part or all of it has to be removed, or else the floor has been severely damaged. In such as case, it can be broken up with the claw end of a hammer, and then shoveled out. It is possible to save the unsealed layer for a future project. To do this, first remove the oiled layer, using a hammer and pry bar to pry and peel it off. Dispose of this part, then remove and store the unsealed material, which can be rehydrated and used again at any time. The oiled layer cannot be rehydrated but can be disposed of in a landfill, or even broken up and spread on the ground, as it contains no toxic ingredients (provided an all-natural sealing oil was used).

SECTION 5

BEYOND THE BASICS

If you've come this far, you've learned how to successfully install your own earthen floor. Congratulations! We hope it was an enjoyable process. If you install more floors, you'll find it gets easier each time. And once you're ready to get beyond the basics, there are many ways to modify the technique, add artistic touches or venture into uncharted territory by trying out some experimental techniques. The possibilities are endless!

(Photos: Left: Mike O'Brien, Right: James Thomson)

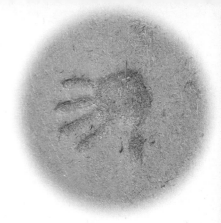

CHAPTER 11

Artistic Touches and Other Techniques

BUILDING IS A CREATIVE PROCESS. Seeing the physical result of one's own imagination and hard work is a special kind of delight. The most beautiful floors are the ones on which the installer has taken time to add their own personal creative touches, to make them truly unique and special. Incorporating artistic details makes the process more enjoyable and satisfying. A floor is like a giant canvas that can be decorated in a variety of ways, such as with tile or stone inlays or prints or paints. There is ample opportunity for budding and seasoned artists alike to express themselves.

Inlays

Tiles, stones and other solid (flat) objects can be embedded in an earthen floor. This can be a wonderful way to define an area, mark an entrance and add texture to a floor.

The challenge with inlays is using the right size and thickness of tile (or stone) so that it

Sukita:

I am fascinated by all types of floors. I see designs I could mimic from old concrete floors and wood floors, and even prints in wallpaper that give me ideas for printing on earthen floors. The simply beauty of a flat floor also delights me. Mixing colors, sprinkling on color, simply being free to try things is exciting! Let us know what you try!

An example of using accent tiles in an earthen floor.

(Credit: Mike O'Brien)

is stable in the floor. A tile that is large and thin will be more likely to pop out or break than a smaller, thicker tile. Generally speaking, if a tile is larger than 4"x4", it should be at least ¼" thick. Smaller tiles can be thinner (down to ⅛"). Consider the location of the inlay, too; a tile or stone placed in a high-traffic area must be more stable and secure than one in an out-of-the-way location.

Unsealed tiles and porous stones can be stained by the wet mix and by the oil sealer. If this is a concern, cover the top surface of the tile with painter's tape before setting it. Be sure to indicate the direction you want the tile to be placed on the tape if this is important. The tape should only cover the top (not the sides) of the tile, so that it is easy to remove once the floor is dry. Tape does not do a great job at protecting tiles from oil, so take it off before oiling and allow the tiles to be oiled, or apply oil carefully so as not to get it on the tile. Unglazed tiles will take on a darker look when oiled.

Top left: *Pigmented floor in Buddist Dance Temple.* (Credit: Mike O'Brien)

Top right:
Rustic floor in Mexico, with donkey manure fiber.
(Credit James Thomson)

Bottom Left:
High gloss finish on an earthen floor in living room.
(Credit Sally Painter)

Top:

*Applying color
wash.* (Credit
Miri Stebivka)

Bottom left: *Powdered pigments.* (Credit: Miri Stebivka)

Bottom right: *An array of color options.*

(Credit: Sukita Reay Crimmel)

Top left: *Color wash before oil.*
(Credit: Sukita Reay Crimmel)

Bottom left: *Color wash in process, note pattern of brush application.*
(Credit Sukita Reay Crimmel)

Bottom right: *Living room with red pigmented earthen floor.*
(Credit James Thomson)

Top: *Steel trowels and wood float, tools for pouring and smoothing.*
(CREDIT: MIRI STEBIVKA)

Bottom left:
Wood float in action, leveling a floor.
(CREDIT: JAMES THOMSON)

Bottom right:
Sukita and Erika having fun in the mud!
(CREDIT: JAMES THOMSON)

Top: *Steel trowel in action, smoothing a floor.* (CREDIT JAMES THOMSON)

Left: *Toys on a matte finish earthen floor.* (CREDIT MIKE O'BRIEN)

Right: *Sukita touches a floor.* (CREDIT MIKE O'BRIEN)

Top: *Earthen floors can work well in kitchens.*
(CREDIT: MIRI STEBIVKA)

Middle left: *Stools!*
(CREDIT: MIRI STEBIVKA)

Bottom left: *Tile, concrete, and earthen floors.*
(CREDIT: JAMES THOMSON)

Bottom right: *4" painted tiles embedded in a floor.*
(CREDIT: MIKE O'BRIEN)

Setting tiles

Small tiles (1" square, and less than ¼" thick) can be simply pushed into the troweled surface of the floor as it is being poured. The surface of the tile should be at or just slightly above the surface of the finished floor.

It can be difficult to push thicker and larger tiles into the mix without disrupting the flat, level surface. Set the tile lightly on the surface of the wet floor and press gently to make a mark. Then lift it off and use a small trowel or even a spoon to create an indentation the same size as the tile. Set the tile into this indentation, vibrating it gently so that wet mix surrounds it on all sides. It is important to not leave any voids under the tile.

Don't set multiple tiles or stones too close together. Earthen floors are good binders, but not nearly as sticky as the thinset mortar and grout used for standard tile installation. Keep the tiles at least half an inch apart.

Once the floor is dry, remove any tape from the tile surface and apply oil over and around the inlay. Be sure to saturate

Small tiles can be pressed into the wet finish. (Credit: James Thomson)

Remove some of the surface material to create a space for a larger tile. (Credit: James Thomson)

An installer leaves her mark. (Credit: James Thomson)

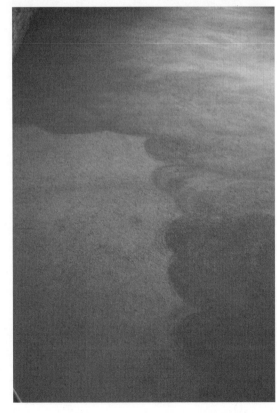

A color wash in process. (Credit: Sukita Reay Crimmel)

with oil any separation that develops between the tile and the earth mix or any hairline cracks running off the tiles.

Note: Inlaying stones and tiles is a time-consuming process. Contractors should make sure to add extra time if their bid includes an inlay.

Prints: Patterns can be stamped into the floor, either by creating a stamp (from a woodblock or piece of metal) or using a found object (a leaf, feather, etc.). The feet and handprints of the builders are also popular choices.

Adding color with color washes, paints and pigmented oils

Applying a color wash

Using a color wash can alter the color of part (or all) of the floor. A color wash is a mix of clay, fine aggregate, water and pigment that is applied to the surface of floor once it has dried but before it is oiled.

The wash should be quite thin; if it is too thick with clay, it will prevent oil penetration. Deep oil penetration is critical to the long-term durability of an earthen floor.

A basic color wash recipe is below. Experiment with different amounts of pigment to achieve the desired result:

> 1 part finely sifted clay (kaolin or site clay sifted through a window screen)
> 2 parts fine mica or sand (300 mesh) (suggested but optional)
> 16 parts water
> pigment to taste (just kidding — don't taste it, keep to one part or less of pigment)

The wash can be applied simply by brushing it on, or by brushing followed with a stiff trowel. Brushing and troweling will leave visible marks in the finish; make tests if aesthetic subtleties are a concern. Keep mixing the wash as it is being applied, as the heavier ingredients will settle down to the bottom of the container.

Painting pictures or designs

Color washes can be used to paint patterns or designs directly onto the floor before it has been oiled. A richer color may be desired for this purpose; add more pigment to the wash to deepen the hue. A

A tree design painted with a color wash.

(Credit: James Thomson)

natural oil-based paint could also be used. The paint should be thin enough that it can penetrate well into the floor, and ideally have a linseed oil or natural drying oil base (see below). See Resources appendix for a suggested supplier.

Pigmenting the oil

Pigment can also be mixed with the linseed oil and applied to the entire surface of the floor when it is sealed. This technique will give a different aesthetic than the color wash described previously. A color wash will often show brush or trowel marks and is more opaque, hiding some of the straw. Pigmented oil can be more translucent, allowing more of the texture of the dried earth to show through. This will vary depending on the type and amount of pigment used; tests are always recommended. It is difficult to give a recipe for an oil–pigment blend because the qualities of pigments vary so widely. The pigment should make up 5 percent or less of the total volume; too much pigment can leave a layer of pigment at the surface that may flake off.

There are two possible advantages to pigmenting the oil instead of doing a color wash. The first is that using it cuts one step — the color washing — from the installation process. The second is that pigmented oil could be used to refinish the floor at a later date, changing the color of an existing floor. As discussed in Chapter 2, earthen floors are limited to fairly dark and earth-toned colors, due to the nature of the mix and the darkening properties of the oil.

Tempera: There is a woman in France, Marie Milesi, who has developed a method of applying a tempera to freshly oiled earthen floors; the result is a very light finished color. Tempera is a paint usually made of pigment and eggs. The details of her process are laid out in her book *Les Sols en Terre* (see References).

Other options

Earthen floors are not new. As discussed in Chapter 1, people have been living on them for millennia, and have developed many different installation techniques that are well-suited to specific materials and environments. The technique outlined in this book has been refined and proven over more than a decade of work, but there are many other options that others have used successfully for generations. These options fall into several categories:

- Radically different techniques
- Different options for the mix
- Different fibers
- Different options for the sealers and finishes

Radically different techniques

One alternative is the "tamped-earth" floor. Instead of making a wet mix and smoothing it out with a trowel, this technique calls for a much drier mix that is tamped into place with a metal tamper. The tamping action compacts the mix into a hard, solid surface and also leaves it flat and smooth.

Alejandra Caballero has been using this technique effectively in Mexico for many years. She finds it works well there because good drying weather is sometimes limited (this technique uses much less water), and they have had trouble with poured floors cracking. This is a labor-intensive process (lots of tamping), but it results in durable and very flat floors.

Different options for the mix

Lime floors: Lime is a hard-setting binder similar to cement. It is made from limestone and is used for plasters, whitewashes and as a cement alternative in mortars. Some builders have experimented with using lime in place of or in combination with clay, to get a harder, faster-drying floor. Some of these floors have been installed in very wet environments, like showers.

The chemistry of clay and lime is complex, and there is not enough data to provide helpful guidelines for making this type of floor. Including (or substituting) lime can create a floor that is lighter in color and a finish that may be more water-resistant than that of a "clay-only" floor. Always make tests. It is still recommended to seal a lime floor with oil and wax.

Different fibers

Any strong natural fiber will likely work in place of chopped straw, as long as it is short and flexible enough to be smoothed out. The most common alternatives to chopped straw are:

- **Hemp:** Hard to get, and must be chopped and sifted. A reasonable option if available.
- **Coconut fiber:** Common in tropical regions, also available in compressed blocks at garden stores. Will require separating and chopping.
- **Fiberglass or poly-fibers:** Plastic and fiberglass fibers

Coconut fiber.

(CREDIT: WIN NONDAKOWIT | DREAMSTIME.COM)

Fiberglass fibers, used in the concrete industry.

(CREDIT: JAMES THOMSON)

Paper fiber. (Credit: Miri Stebivka)

A floor made with paper fiber. Note the absence of visible fibers (straw).
(Credit: Mike O'Brien)

manufactured specifically for use in concrete work will also work with earthen floors, but these fibers do not bend well and may stick out of the floor. And they are not made from natural materials.

- **Paper fiber:** Paper fibers are not very strong individually, but enough of them together can add a lot of strength. Adding shredded paper, which is almost like dust, or soaked and blended newspaper, creates a finish that can look like concrete, with no visible fiber.

Shredded wood fiber. (Credit: Miri Stebivka)

Horse manure is an excellent fiber source. (Credit: James Thomson)

- **Wood fiber:** Similar to paper, wood fiber is short and not very strong, but if enough is added, it will work. This type of fiber is a by-product of the manufacturing processes for certain wood products, such as OSB (oriented strand board) and Faswall blocks (a wood-chip-cement block). During the pouring and troweling process, be cautious of the little fibers sticking out of the mix, as they can cause splinters. Burnishing will push these down and prevent further splinters. Wood fiber is visible, but is smaller than most straw fibers.
- **Manure:** Dried manure from horses, cows and other large herbivores makes for a great fiber option. These animals eat fibrous plants, breaking them down with their teeth (and digestive tracts) into small pieces that are perfect for floors and plasters. Other ingredients in manure give it a bit of a "binding" quality as well, which improves the stickiness and workability of a mix (similar to adding more clay, but without the risks of shrinking). Dried horse manure screened

through an ⅛" screen is the preferred option because the fibers are longer; if this resource is available it is highly recommended. Often it is used along with other fibers (e.g., chopped straw). Another great advantage to manure is that any seeds have been digested, so there is little risk of sprouts in the floor. People are sometimes squeamish about the thought of working (and living) with manure. The reality is that it's a very pleasant material to work with; any minimal odor in the raw material vanishes once the floor is dry.

Different options for the sealers and finishes

Along with the solvent, the sealing oil is the most expensive ingredient in an earthen floor and requires the most careful handling. Many people have been trying to figure out an effective substitute. The reality is that there are few other good options.

Drying oils: There are other drying oils (see Chapter 3) that can be used. Boiled linseed oil is the most common choice and has the most successful track record. Other drying oils include:

- Walnut oil
- Tung oil
- Poppyseed oil
- Perilla oil
- Sunflower oil
- Safflower oil

Which oil to use will ultimately depend on what is most available and cost effective. Linseed oil makes the most sense for many people. Oils that are not intended specifically as a wood (or floor) treatment may take a very long time to dry. Make tests when using unfamiliar oils.

Ox blood: There is a tradition of using blood as sealer for earthen materials in some parts of the world, including the southwest of the United States and Mexico. This technique is not in common

Peggy Frith's peanut-oil floor:

For my first earthen floor, we made a mix that had a lot of manure and clay, it cracked a lot. It dried well but had so much manure in it that it was quite porous, almost spongy. I didn't have any money at the time, so there was no way I could afford the linseed oil. I went and bought a five-gallon bucket of peanut oil because I read in one of those "old-timey" books that people used to seal floors with it (I've never been able to find that book since!).

I had just learned how to oil floors from Rob Boleman. He taught that you needed to thin the oil with solvent before applying it. We managed to scrounge ten bottles of Citra-Solve, which is essentially citrus oil. We mixed the Citra-Solve and the peanut oil and applied about eight coats — the floor just kept absorbing oil.

It took a really long time to dry — like months. At first it was so soft you could put prints in it. It's funny, I was totally confident that it would work; I had no idea what I was doing! It was at least two weeks before we started walking on it, and then only with socks. The surface stayed oily and soft for weeks. I think it finally lost its stickiness because dust stuck to the sheen of peanut oil.

Later when I moved out of the house, I had to get my woodstove out. It was a cast-iron stove, very heavy with pointy legs. I didn't have anyone there to help me, so I dragged it across the floor. It left grease skid marks but didn't scrape the floor at all!

practice today in North America; it is impractical (just finding a source of fresh blood would be difficult), and if using manure is a tough sell, ox blood would probably be a "deal breaker" for most people. Blood coagulates as it dries; this is a similar process to the polymerization of linseed oil, leaving a floor that is water resistant and dust free.

Options for wax

A finishing wax is not necessary, but many choose to apply one to increase durability and for enhanced luster (shine). The Resources appendix includes the recommended brands for earthen floors.

These waxes have different ingredients and produce different results. The Bioshield Floor Finish, for example, is a soy resin/

carnauba wax finish that leaves a very shiny surface, while the Claylin Wax leaves a matte finish. Others have experimented with conventional wood floor wax products and have reported good results. Consider the ingredients of the wax; request MSDSs if possible.

It is possible to make a wax blend at home. This is an area for further experimentation; try a local bookstore or the Internet for recipes. Ingredients to these recipes typically include:

- Oil (olive, walnut, linseed)
- Beeswax
- Carnauba wax
- Solvent (alcohol or citrus-based)

Homemade waxes are typically thicker than the recommended waxes (see Resources appendix), so will need to be rubbed on with a rag. It is possible to increase the oil and solvent content in the

Making a homemade floor wax blend on a stove top.

(Credit: James Thomson)

wax to make it thin enough to apply with a brush or woolie pad, as described in Chapter 9. The danger with a homemade wax is that it may be applied too thick and not buffed off properly; this can leave a sticky waxy film on the floor. Using an electric buffing tool to spread and remove excess wax is strongly recommended (see Chapter 9).

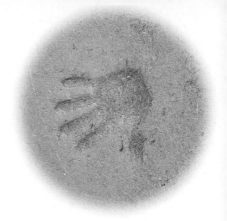

CHAPTER 12

Floor Experiments

Using burlap

BURLAP FABRIC, FROM OLD COFFEE SACKS or rolls used in
landscaping, can be a helpful addition in specific situations.
For example, the recommended minimum thickness for an earthen
floor over a wood-framed subfloor is ¾". There are some situations
where a thinner floor may be desirable. James encountered this in
his house, and used burlap sheets to reinforce the floor.

Burlap could be used anywhere that additional reinforcing is
desired. Examples are directly over a concrete slab that has cracks
or expansion joints (although a vapor barrier in this situation would
serve the same purpose), or over existing tiles, such as vinyl, ceramic
or asbestos. Other areas to use burlap could be in high-traffic areas,
corners or door thresholds.

Scratch patterns

When the floor is still damp, it can be scratched with a pattern, like
brush-finished concrete. The fibers in the earthen floor mix may
make this a bit challenging, but by adjusting the mix or making
larger scratches, some interesting aesthetic effects could be achieved,
or a textured "no-slip" surface. Keep in mind the challenges of
cleaning deeply scratched areas.

A scratched pattern in a concrete floor.
(CREDIT: JAMES THOMSON)

James:

I was remodeling a bedroom on the second floor and wanted to install an earthen floor. This was a 1926 Craftsman-style bungalow in Portland, OR, and I was concerned about whether it could support the weight of the floor. There was a bit more deflection in the subfloor than I usually like to see for an earthen floor installation. Additionally, the headroom in the second-floor bedroom was very limited. Adding even another half-inch in floor thickness would have made a significant impact in the feeling and functionality of the room. ☞

Plywood subfloor with taped seams ready for ultra-thin earthen floor. Notice the burlap fabric in the center. (CREDIT: JAMES THOMSON)

I wanted to try a floor that was very thin (½") and would be able to withstand higher-than-normal rates of movement without cracking. The room was small, about 10' x 10', and as a bedroom would not be a heavy-traffic area, though it was still possible that movement and bounce would cause the floor to crack over time.

I had the idea of embedding a layer of loosely woven burlap cloth in the floor, to create a strong, interconnected floor with lots of tensile strength.

The subfloor was 2x6 joists, 16" on center. On top of that, I placed ¾" tongue-and-groove plywood (standard subfloor material). As a further experiment, I decided against using a vapor retarder to protect the plywood from the wet mix. My thinking was that because the earthen floor was so thin, it would dry very quickly and present little risk of damaging the plywood. The biggest concern was that water could seep between the seams in the plywood and saturate the sides of the plywood, which could take much longer to dry and potentially cause warping. To eliminate this possibility, I taped the seams between the 4x8 sheets of plywood with duct tape.

On top of the plywood, we first laid down ¼" of wet floor mix, then rolled out and smoothed down the burlap and covered it with another ¼" of floor mix. After smoothing the surface, we moved on to the next section.

The floor dried in a couple of days, and was sealed with oil and wax within two weeks. Completed in December 2010, it has performed very well with very little cracking (just a few superficial cracks that appeared immediately after drying). This experiment can be claimed a success, though it should be noted that I cannot be sure the burlap was necessary. It is very possible that a very thin floor would perform well in this situation even without a burlap layer. I thought about doing half the room with burlap and half without, but ultimately decided it was not worth the risk.... why not do everything possible to reduce the likelihood of cracks? It would be good to do further experiments with very thin floors without burlap to see the results.

James spreads the burlap fabric over the first thin layer of floor mix. (CREDIT: SUKITA REAY CRIMMEL)

Reducing sprouts

Sprouted seeds can be a real hassle, especially if there are a lot of them. Experiments are needed to help identify ways to reduce sprouting. One interesting possibility is to mix a naturally occurring herbicide into the mix. Walnut trees, for example, secrete a chemical (hydrojuglone) from their roots that prevents plants and shrubs from growing up around their roots. Perhaps this chemical, or some natural variant, would be effective in reducing sprouting in an earthen floor. Mixing in borax, a naturally occurring mineral, could achieve the same result. Adding another element to a floor mix may have unpredictable effects on the floor's appearance and durability, so careful experimentation is necessary in this area.

An earthen floor transitions into a plaster.

(CREDIT: MIKE O'BRIEN)

Eliminating VOCs

Mixing solvents with drying oils is a practice that comes from the wood industry, because many hardwoods need the solvents to help the oil penetrate. With earthen floors, it appears that the oil penetrates fairly easily, so it may be possible to use the oil straight, with no solvent. Gentle heating of the oil can help thin it too, and aid in penetration. If successful this would eliminate one of the more dangerous ingredients from the process, the solvent, which contains VOCs.

Bringing the earthen floor up the wall

Most of the time earthen floors stop where they meet the wall.

But why end there? The floor material could be continued up the wall to create an "earthen baseboard" effect, or even further up the wall to make "earthen wainscotting" (see photo in Chapter 2). Something to keep in mind with this idea is the clay content of the mix. Floor mix has a fairly low clay content, but plaster on walls needs more clay to stick to the vertical surface. The solution is to increase the clay content for the vertical section. The wall may also need to be prepared with lath or another adhesion technique. As always, do tests before the actual installation.

Countertops

Concrete countertops have become very popular, why not earthen countertops? The same technique used for pouring a floor can be adapted to make a countertop. There are a few challenges to overcome:

- An earthen floor tends to have a fairly porous surface, which may not be suitable for a countertop that frequently gets wet or is used for preparing food. The mix for a countertop needs to have a smaller range of particle sizes to create a tighter surface. Use sands graded 80 down to 300, and add calcium carbonate, also known as *whiting*, to provide some very fine particles. Screen clay soil through window screens or use a purchased dry clay such as kaolin or another bagged pottery clay.

An earthen countertop installed by Erik Bowden of Island Village Builders.
(CREDIT: ERIK BOWDEN)

- Countertops are exposed to more water than floors, so it would be wise to make sure the entire thickness of the layer gets saturated with oil. This could be done by mixing oil into the mix or by pouring the layer about ½" thick.

- An earthen countertop will not have the same tensile strength as a concrete countertop, so it will have to be fully supported underneath and at the edge by wood or metal.
- The subsurface needs to be rigid with no flex. Before applying the mud mix, install either a mechanical attachment of metal lath or paint on a highly textured sand and paint/primer adhesion coat.
- Countertops need to be very smooth, while a typical earthen floor is more textured and rough. One solution to this would be to sand the surface after oiling with a 400- or 800-grit diamond polishing pad (designed for polishing concrete). This technique would be prohibitively time-consuming for a large floor, but has shown to add a high level of shine and polish to an earthen surface.

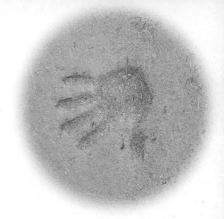

CHAPTER 13

Using Radiant Heat

"R ADIANT HEAT" defines a heating system that heats with radiant energy, as opposed to convective or conductive energy. Common radiant heating systems include baseboard and traditional steam radiators, wood-burning stoves, masonry heaters and the sun. A common convective heat source is a forced hot-air (duct) system.

In recent years, many heating specialists in North America have started to install radiant heating systems in floors, walls and even ceilings (this has been a common technique in parts of Asia and Europe for years). The internal air temperature inside a radiantly heated room or building may actually be lower than in a conventionally heated building without effecting the comfort level of the inhabitants, because people feel the radiant heat directly on their bodies. People report that they love the feeling of a warm floor underfoot and say that it makes the rest of them feel warmer even though the actual room temperature may be cooler. Another advantage to radiant heat is that no air is circulated through (and between) rooms, which results in less spread of airborne particles (aka dust).

The choice of whether (and where) to install radiant heat goes beyond the scope of this book, except to say that earthen floors

are well-suited for under-floor radiant heating systems because they provide a lot of thermal mass.

Radiant floor heat options

There are two commonly used kinds of radiant floor heat. Experiments have been done with other techniques, two of which will be mentioned here.

The most common method is to use hot water to distribute the heat (called "hydronic heating"). Plastic water pipes are buried into the floor and connected to a water heater or furnace and pump. These systems can be installed by a savvy DIYer, but most homeowners choose to have the installation done by a professional plumber. Professional installers will be familiar with the requirements of this system, which include floor insulation, a metal grid for tying the pipe on to and many feet of durable, half-inch plastic pipe

An earthen basecoat poured over radiant hydronic heat tubes. (Credit: Sukita Reay Crimmel)

(PEX or Wirsbo are common brands). The insulation layer is laid down first, followed by the metal grid. The plastic pipe is attached to the grid using twist ties or zip ties in a continuous looping pattern, and is then buried in the earthen floor.

The amount of earthen material on top of the tubes will depend on the system design. In all designs, there needs to be at least one inch of material above the height of the tubes, so the minimum total thickness of the floor will be the thickness of the tubes plus an inch. In the Pacific Northwest and other mild climates, this is as deep as the tubes usually go. This is because the heating needs are less than in colder climates and most people want a quicker response time to feel the heat. Colder climates might warrant more mass, to heat up in the fall and keep warm all winter. For anything thicker than two inches, two layers should be poured (basecoat and finish coat: see Chapter 6). Consult with an expert to determine the recommended thickness of the floor. Most installers will not be familiar with the specifics of earthen floors, but the system specifications can be designed as if a concrete floor is being used.

When pouring an earthen floor over hydronic tubes, it is important not to damage the tubes. They are made of a very strong material designed to withstand a lot of weight, but why risk it? Precautions can include pressurizing the system with water or air and laying plywood along the delivery path so the wheelbarrow will not roll directly on the tubes. When lifting up the wheelbarrow to dump material, be especially cautious that the front metal bumper is not pivoting on any of the tubes.

Another type of radiant heat system uses electric heating elements. These come in rolls or mats and are usually intended to

A pressure valve attached to a hydronic tubes system so it can be pressurized before installation of the earthen floor. (CREDIT: RON HAYS)

go under tile, wood or laminate floors, but can be buried under an earthen floor. Electric radiant heat is simpler and cheaper to install, but may end up being more expensive in the long run, as the cost of electric heat is usually higher than the cost of operating a gas- or oil-fired water heater or furnace. The pads themselves are quite thin, which means the overall floor thickness can be less than it would be when using hydronic heat. Consult with a professional installer to determine the appropriate thickness of the floor, making sure to keep it to at least ¾" (the recommended minimum for any basic installation).

A more experimental and less common option is to use hot air to heat a floor. One technique involves running the exhaust from a wood-burning "rocket stove" under a floor; the hot exhaust (smoke) releases its heat into the mass of the floor, which then slowly heats the room. Another approach is a passive conduction system, called a hypocaust. This design is inspired by an ancient Roman technique in which warm air (heated by the sun or a wood fire) circulates beneath the floor and warms the mass in the floor and the room above. These are experimental techniques and should only be attempted by those who are interested in being pioneers in this area! Learn more about rocket stoves in the book *Rocket Mass Heaters* by Ianto Evans and Leslie Jackson (details in the References appendix).

A rocket stove, with the exhaust being installed under the subfloor.

(Credit: James Thomson)

CHAPTER 14

The Future of Earthen Floors

HOMEOWNERS AND BUILDING PROFESSIONALS are hungry for more environmentally friendly ways to build. The green building industry has seen rapid growth in recent years, despite the serious recession. The number of architecture and construction firms undertaking "green" projects has increased from 13 percent in 2009 to 28 percent in 2012, and is expected to reach 51 percent in 2015.[7] This is in large part due to the popularity of green building benchmark programs like LEED[8] and the Living Building Challenge.

Earth building and natural building are tiny niches within the green building industry, so there is plenty of room for growth. The Living Building Challenge[9] sets strict limits on how far materials can travel from their point of origin to the building site. For heavy materials (earth, sand, concrete), the limit is about 350 miles. Earthen floors, and many natural building materials, fall easily into this category, as in most places the raw ingredients (sand, clay) are available very nearby.

Because earthen floors can be installed in any home and use installation techniques that are similar to the concrete techniques many builders are familiar with, they have the potential to become popular choices for sustainability-minded homeowners.

Natural building has typically been the realm of the DIY home-owner and/or homesteader. People are drawn to the techniques for philosophical reasons as much as anything. These techniques will always be appealing to those who like to "do it themselves," because the materials are safe, inexpensive and just fun to work with. But that doesn't mean there can't also be a healthy commercial market, with professional contractors installing floors for homeowners who would prefer to pay someone else rather than do it themselves. There are already a few practitioners scattered around the country who offer this service. Increased awareness will increase demand, creating room for more builders with the skills and knowledge to install earthen floors professionally. Existing contractors may see value in adding this technique to their repertoire, as a way to attract homeowners who are looking for truly sustainable building options.

> US demand for hard surface flooring is forecast to in-crease 5.6 percent per year through 2015 to 8.6 billion square feet, with a value approaching $11 billion in the United States.[10]

If earthen floors could capture only one-tenth of 1 percent of this market, that would be $11 million ... nothing to sneeze at!

Running a natural building contracting business

There are certain challenges in being a natural building contractor. The fast-moving building industry relies on systems and products that have known and consistent results. Contractors rely on this consistency to bid competitively for jobs. The challenge of working with local natural materials is that they are inherently inconsistent; the characteristics of a clay from one region will vary from that of another (even neighboring) region. Contractors must explain to clients that earth-based materials are different from most other conventional options. Colors and textures may vary from the samples, the floors will wear and change over time, and typically won't be perfectly flat and smooth in the way most expect finishes to be. These qualities are what make earthen floors (and other earthen finishes) so beautiful and unique, but it's important that potential clients understand this from the start.

Most people who engage in professional natural building work are not simply trying to make a living; they are also trying to

Sukita: The story of Claylin

After years of creating recipes for earthen floors, I decided to start a manufacturing business that mixes and bags an earthen floor recipe I had become familiar with. Having a single-source product allows for more consistent results, and the fifty-pound bags make it possible to deliver the exact amount of material needed for each floor. At first I looked into buying the tools needed to screen and bag large amounts of earth and sand and found the upfront costs would be quite large. Instead I partnered with another business that already does this type of thing, a company that mixes clay soils for sports fields. After months of testing and development, I now have a consistent product that can be used in any floor installation. I call the company Claylin, a combination of the words "clay" and "linseed oil." Claylin produces a mud mix as well as the fiber, oil and wax needed to make a floor. These products have reduced many of the unknowns, leaving room to focus on education and promotion. Currently Claylin manufactures in Portland and markets in Oregon, Washington and Northern California. Future growth plans include manufacturing hubs in other regions, to provide materials that are as local as possible.

Claylin product line.
(CREDIT: JAMES THOMSON)

incorporate their values into their livelihood. Natural businesses can be managed with many of the same philosophies that go into natural building. In *The Ecology of Commerce*, author Paul Hawken, suggests three principles for doing business this way:

> The first is to obey the waste-equals-food principle and entirely eliminate waste from our industrial production. The second principle is to change from an economy based on carbon to one based on hydrogen and sunshine. Third, we must create systems of feedback and accountability that support and strengthen restorative behavior, whether they are in resource utilities, green fees on agricultural chemicals, or reliance on local production and distribution. All three recommendations have a single purpose: to reduce substantially the impact that each of us has upon our environment. We have to be able to imagine a life where having less is truly more satisfying, more interesting, and of course, more secure.[11]

For a contractor who is paid by the hour, the time required to do new tests for every new job can be prohibitive. There are a few companies in North America (and others abroad) that produce earth-based building products, like paints, plasters and even bagged earthen floor mix, to help reduce some of the challenges associated with sourcing, mixing and providing consistency of materials and costs. Hopefully these products will make earthen building options more accessible and feasible for homeowners and contractors alike.

Conclusion

WORKING WITH EARTH IS A DELIGHTFUL EXPERIENCE. As with any building technique, there are a lot of details to keep track of in order to increase the likelihood of a successful installation. This book aims to provide as much information as possible about earthen floors, information gathered and discovered over

An alter on an earthen floor.

years of learning and trial and error. By this point, some readers may be feeling overwhelmed by the all the details, techniques and mistakes to avoid. Do not be discouraged; no one can become an expert floor installer simply by reading a book.

The reality is that earthen materials are incredibly forgiving and a pleasure to work with. The best way to learn how to use them is to simply get started. Dig up some soil, find some sand, mix it up, add some fiber and start playing. Get to know the materials and enjoy the process. You will make mistakes; the authors are still making them after years of experimentation. But through experimentation comes discovery, and there are many creative applications yet to be discovered. In the small-but-growing natural building field, people are always finding interesting ways to use the materials, adding their own unique artistic touches and making technical improvements and efficiencies. There are many floors still to be troweled, oiled and walked on. With luck, the tools in this book will play a role in encouraging more and more people to bring some earth into their home.

Sukita:

I am always so heartened by the positive response I get to earthen floors. I visit many "green building" home shows to promote my work; these shows are always filled with an overwhelming array of the latest and greatest green building materials. It's inspiring to see how, among the countless displays and products, people are drawn to the earthen floor samples I show; they are like beacons of light in the chaos.

In December 2012, I entered earthen floors into the Oregon BEST Red List Design Challenge, a competition to develop Oregon-made building materials that meet the Living Building Challenge protocol. In May 2013, I was selected as a finalist, and I presented my flooring product at the Living Building Institute's unConference, a gathering of about a thousand like-minded builders, designers and architects. In September, I learned that I won first place in the contest! It was very exciting to see earthen floors acknowledged as a promising option for sustainability-focused building projects and professionals. This type of exposure is important to help spread the word about and grow the market for earthen floors.

James:

Even after years of experience building with earth, I sometimes still find myself awestruck by the technique. I would find it hard to believe that it is possible to mix together a few simple ingredients — ingredients that can be found right beneath my feet, or on a hillside nearby — and build a house, plaster a wall or make floor, if I hadn't seen it happen so many times.

I love working with earth because it is forgiving, flexible and fun. It's not the fastest way to build, nor necessarily the easiest, but it is satisfying beyond any other technique I have tried. Mixing mud and spreading it flat and smooth is an almost meditative process, one that creates a primal connection between the world around me and the space I shelter myself within.

A floor is such a basic element of a home: so important, yet often unnoticed. We cover our floors with rugs and furniture and let dust gather on their corners. But a floor is literally and figuratively the foundation we build our homes and our lives upon. To have a floor made of earth is a real treat, especially when it's one you've made yourself. I'm not much of a mystical type, but there is something magical here, something powerful in taking the simple elements around us and transforming them into a durable surface to live life upon. I hope readers of this book will try their hands — and feet — at this remarkable technique.

The authors, relaxing with their favorite tools.
(Credit: Miri Stebivka)

Appendix A:

Glossary

¾" minus: A sifted and washed mix of large and small pieces of gravel that can pass through a ¾" screen. This material is used for making stable bases and pads for building; it tamps down very hard, especially when using a plate compactor.

Aggregate: A component of a composite material that helps to resist compressive force. Usually rock or mineral fragments.

Cob: A monolithic earthen building material and technique originating in the British Isles made from clay soil, sand and plant fiber. This ancient technique has been adapted and improved for modern use by pioneering natural builders in Oregon, and is sometimes referred to as "Oregon cob."

Cold Joint: Describes an area where fresh wet mix is laid down next to material that is older and perhaps drier. If this area is not blended well, a crack will develop along the joint.

Compressive strength: The ability of a material to withstand a load against it that reduces its size (by squeezing). In construction, this typically applies to materials that support the load of a building, like the foundation or the vertical structural members in the wall.

Concrete pad: A slab made from concrete, usually "on-grade." Common in basements, ground floors, garages, etc.

Control joint: A cut made in a monolithic slab material that is intended to control where cracks occur in the material. A control joint typically cuts through half or three-quarters of the layer.

Clay slip: A suspension of clay in water.

Drainage rock: Large, uniform-sized gravel used to fill drainage trenches or other areas to aid in drainage and movement of water. Typically round, about ¾" to 1½" in diameter. Also referred to as washed rock, river rock, drain rock.

Expansion joint: A split or break that goes all the way through a monolithic slab (concrete, earth, etc.) and extends along the entire width of the slab. Intended to break large expanses up into smaller sections that are less likely to crack with movement and expansion/ contraction.

Float: A tool with a very thick, non-flexible wood or magnesium blade. Used to roughly smooth the surface of an earthen mix before smoothing with a steel trowel. Leaves a surface with an "open," porous texture.

Grade: The level of the ground.

Hardware cloth: Metal screen material that comes in a range of mesh sizes, from ⅛" to ¾". Comes in rolls of varying widths. Used for screening raw ingredients.

Hydronic radiant heat tubes: Plastic tubes that carry hot water through a high-thermal-mass material (concrete or earth) to warm the material and radiate heat into a room. Part of a radiant heat system.

Level: *adjective* — Exactly horizontal.

MSDS: An abbreviation for "Material safety data sheet." An informational sheet provided by manufacturers of products that contain hazardous materials to assist workers and emergency personel with the safe handling of those materials.

On-grade: Built directly on the ground.

Perm: A unit of "permeance" or "water vapor transmission" through a particular membrane.

Plumb: *adjective* — Exactly vertical.

Pool trowel: A trowel with rounded corners and a flexible blade. Used for troweling curved surfaces.

Road base: A well-graded material consisting of coarse and fine aggregate (sand and gravel). Produce by crushing rock or recycled materials (concrete). May also contain fine particles such as silt or clay. Used to create stable bases for roads and buildings. Similar to ¾" minus, but usually not sifted or washed.

Screed board: A wooden board used to move/level material between screed rails to achieve a specific and consistent thickness of material.

Screed guide: Synonymous with "screed rail."

Screed rails: Guides (usually wood) used to support the screed board and determine the thickness of the material being screeded. Can be set on the subfloor if level, or attached to walls to make a level surface on a non-level subfloor.

Screeding: The process of using a screed board and screed rails to flatten a section of earth or concrete.

Slab: A large, flat piece of solid material. Usually in reference to concrete.

Spirit level: Uses a bubble in a liquid-filled glass tube to indicate level and plumb. Also called a "bubble level."

Trowel: A tool with a thin, somewhat flexible stainless or carbon steel blade, used for spreading and smoothing setting material such as concrete, mortar, stucco or plaster. Used by floor installers to seal up the open surface that the float leaves.

Vapor barrier: A plastic membrane that prevents the transmission of water vapor. Defined as a layer with a perm rating of 0.1 perm or less. 6-millimeter black plastic polyethylene sheeting is recommended for most applications.

Vapor retarder: A water-resistant membrane that slows but does not entirely prevent movement of water vapor. Defined as a layer with a perm rating of 0.1 to 1. Applied to plywood subfloors as preparation for an earthen floor installation, to prevent the plywood from becoming saturated with water during the installation process.

Appendix B:

References

CHAPTER 1

Earth Architecture
Ronald Rael
Princeton Architectural Press, 2009
Architecture 2030 (website and organization)
architecture2030.org/the_problem/buildings_problem_why

The Elements of Style: A Practical Encyclopedia of Interior Architectural Details
Stephen Calloway and Elizabeth Cromley
Simon & Schuster, 1997

CHAPTER 2

State of California Environmental Protection Agency's updated list of chemicals known to the State to cause cancer or reproductive toxicity:
oehha.ca.gov/prop65/prop65_list/files/P65single052413.pdf

Inventory of Carbon and Energy
Geoff Hammond and Craig Jones

Department of Mechanical Engineering
University of Bath, 2006

Adobe and Rammed Earth Buildings
Paul McHenry
University of Arizona Press, 1989

CHAPTER 3

Richter, D.D. and Markewitz, D., *Bioscience* 1995, 45: 600-609, and
Paul, E.A. and Clark, F.E., *Soil Microbiology and Chemistry*. New
York: Academic, 1989

The Straw Bale House
Bill and Athena Steen
Chelsea Green Publishing, 1994

Clarifying raw linseed oil
fullchisel.com/blog/?p=1611

Color: A Natural History of the Palette
Victoria Finlay
Random House, 2003

CHAPTER 13

Rocket Mass Heaters: Super Efficient Woodstoves YOU Can Build
Ianto Evans and Leslie Jackson
Cob Cottage, 2006

CHAPTER 14

World Green Building Trends SmartMarket Report, McGraw Hill
Construction (2013)
construction.com/about-us/press/world-green-building-trends-
smartmarket-report.asp

Earth Advantage Institute
earthadvantage.org

Living Building Challenge
living-future.org/lbc

Summary of "Hard Surface Flooring," a Freedonia industry study:
reportsnreports.com/reports/117994-hard-surface-flooring.html

The Ecology of Commerce
Paul Hawken
Harper Business, 2010

OIL AND WAX SAFETY APPENDIX

Highrise Office Building Fire One Meridian Plaza Philadelphia,
Pennsylvania
Gordon J. Routley; Charles Jennings; Mark Chubb
Report USFA-TR-049, Federal Emergency Management Agency,
1991
usfa.fema.gov/downloads/pdf/publications/tr-049.pdf

Appendix C:

Resources

Oils and waxes

Claylin
Portland, OR
Phone (503) 957-6132
claylin.com

Claylin Oil
Oil and solvent blend designed
specifically for finishing earthen
floors

Claylin Wax
Wax and oil blend designed for
earthen floors

Heritage Natural Finishes
2214 Sanford Dr., Unit 8
Grand Junction, CO 81505
Phone (541) 844-8748

Toll-free 888-526-3275
heritagenaturalfinishes.com

Concentrated Finishing Oil
Oil and solvent blend suitable
for earthen floors

Liquid Wax End Sealer
Wax, oil and solvent blend suitable for earthen floors

Pure Citrus Solvent

Tried & True Wood Finishes
14 Prospect Street
Trumansburg, NY 14886
Phone (607) 387-9280
triedandtruewoodfinish.com

Danish Oil
Refined and heat-treated
(polymerized) linseed oil. Will
need to be mixed with a solvent

to thin it for use on an earthen floor.

BioShield Paint Company
(Products are made in Germany, sold out of New Mexico)
Plaza Entrada
3005 S. St. Francis Suite 2A
Santa Fe, NM 87505
Phone (505) 438-3448
bioshieldpaint.com
Bioshield produces a large line of sealing oils. The following are suitable for earthen floor finishing:

Penetrating Oil #5
A good all-around floor sealing oil

Primer Oil #1
A good all-around floor sealing oil

Hardening Oil #9
Can be applied after the penetrating or primer oils. Helps fill voids and smooth the finish.

Hardwax #32
A wax blend

Resin Floor Finish #4 (Semi-gloss)
A wax-resin finish to apply after oiling, creates a glossy finish

Citrus Thinner #23
For cleanup and thinning the sealing oil if it is too thick

Floor soaps

Ecover
(Manufacturer of natural cleaning products, based in Belgium)
ecover.com

Floor Soap with Linseed Oil
Linseed oil-based floor soap

Bioshield
bioshieldpaint.com

Floor Soap #58
An all-purpose oil-based floor cleaner

Floor Milk #59
A floor polish

Colgate-Palmolive
colgatepalmolive.com

Murphy's Oil Soap
Widely available household floor cleaner, suitable for wood finishes

Pigment

Ochres & Oxides
Phone (707) 243-2252
ochresandoxides.com

The Earth Pigments Company
P.O. Box 1172
Cortaro, AZ 85652-1172
Phone (520) 682-8928
earthpigments.com

Natural paints

Earth Paints LLC
PO Box 94
Ashland, OR 97520

Phone (541) 708-3702
naturalearthpaint.com

Earth Oil Paint Kit
Natural pigments with almond-oil base, appropriate for painting directly on earthen floors before sealing

Vapor retarders

Fortifiber Building Systems Group
300 Industrial Drive
Fernley, NV 89408
Phone toll-free 800-773-4777
fortifiber.com

Aquabar "B"
Moisture vapor retarder for use in a variety of interior construction applications. Consists of two layers of kraft paper laminated with asphalt.

W.R. Meadows, Inc
P.O. Box 338
Hampshire, IL 60140
Phone toll-free 800-342-5976
wrmeadows.com

Red Rosin Paper
Multi-purpose building paper made from recycled fibers. Rosin adds moisture resistance.

Wood fibers

Waste fibers from Faswall, an insulated wood-chip cement form building block:
faswall.com

TOOLS

Oil sprayers

Chapin International
700 Ellicott Street
Batavia, NY 14021-0549
Phone toll-free 800-444-3140
chapinmfg.com

Industrial Viton® Concrete Open Head — model # 1949

Trowels

Marshalltown
104 South 8th Avenue
Marshalltown, IA 50158 USA
Phone (641) 753-5999
marshalltown.com
Manufacturer of a wide variety of trowels, floats, and masonry tools

12" x 3" Finishing Trowel w/ Curved DuraSoft® Handle
Steel trowel for finishing and burnishing. Available in different lengths.

12" x 3 ½" QLT Xtra-Hard Wood Hand Float
Wood float for spreading and flattening. Available in different lengths.

14" x 4" Fully Rounded Finishing Trowel w/Curved DuraSoft® Handle

Steel trowel with rounded corners, good for beginners and hard-to-smooth areas. Available in different lengths.

Japanese trowels

Thin flexible trowels, good for details and hard-to-reach spots; available from:

LanderLand
62 Kingston Main Street
Hillsboro, NM 88042
Phone (575) 895-5029
landerland.com

Flexible Stainless Steel Pointed — Shiage Gote (#17025)
Available in several sizes; 210 mm or 225 mm are recommended.

Trowel oil — Japanese camellia oil (Hamono Tsubaki)

This highly refined camellia oil has been produced from the same recipe for more than eighty years in Japan. It is used to protect saw blades, swords, knives, plane irons, chisels and many other metal objects from corrosion.
Available from:
fine-tools.com/pflege.htm

Insulation

Rigid foam products appropriate for on-grade applications:

Owens Corning
insulation.owenscorning.com

Foamular XPS
4' x 8' extruded polystyrene sheets. Comes in a variety of thicknesses and compressive strengths. Select one with a strength of at least 25 psi.

Dow Chemical
building.dow.com

Styrofoam
4'x8' extruded polystyrene sheets. Comes in a variety of thicknesses and compressive strengths. Select one with a strength of at least 25 psi.

Perlite

Supreme Perlite Company
4600 N. Suttle Rd.
Portland, OR 97217
Phone (503) 286-4333
perlite.com

Bagged Perlite
Suitable for under-slab insulation.

BOOKS

Dirt: The Ecstatic Skin of the Earth
William Bryant Logan
W. W. Norton & Company, 2007

Clay: The History and Evolution of Humankind's Relationship with Earth's Most Primal Element
Suzanne Staubach
Berkley Books, 2005

The Solar House: Passive Heating and Cooling
Daniel D. Chiras
Chelsea Green Publishing, 2002

Passive Solar Simplified: Easily Design a Truly Green Home for Colorado and the West
Thomas P. Doerr
CreateSpace Independent Publishing Platform, 2012

Concrete at Home
Fu-Tung Cheng
Taunton Press, 2005

Shellac, Linseed Oil, & Paint: Traditional 19ᵗʰ Century Woodwork Finishes
Stephan A. Shepherd
Full Chisel Press, 2011

General Natural Building Titles

The Hand-Sculpted House: A Practical and Philosophical Guide to Building a Cob Cottage
Ianto Evans, Michael G. Smith and Linda Smiley
Chelsea Green Publishing, 2002

Building with Cob: A Step-by-Step Guide
Adam Weismann and Katy Bryce
Green Books, 2006

The Natural Building Companion: A Comprehensive Guide to Integrative Design and Construction
Jacob Deva Racusin and Ace McArleton
Chelsea Green Publishing, 2012

The Art of Natural Building: Design, Construction, Resources
Joseph F. Kennedy, Michael G. Smith and Catherine Wanek (editors)
New Society Publishers, 2001

EcoNest: Creating Sustainable Sanctuaries of Clay, Straw, and Timber
Paula Baker-Laporte and Robert Laporte
Gibbs Smith, 2005

FLOOR INSTALLERS, CONSULTANTS AND WORKSHOP TEACHERS

The following is a partial list of builders and teachers that the authors know to have experience working with earthen floors. This should not be construed as a recommendation for any particular practitioner. Available services will vary with each individual or organization.

Organization/person	Location	Website/contact info
Aprovecho	Cottage Grove, OR	aprovecho.net
Bernhard Masterson	Portland, OR	bernhardmasterson.com
Canelo Project Bill and Athena Steen	Elgin, AZ	caneloproject.com
Chad Tate	Crestone, CO	cmtater@gmail.com
Clay Sand Straw Kindra Welch	Austin, TX	claysandstraw.com
Clayin	Portland, OR	claylin.com
Day One Design Erica Ann Bush	Eugene, OR	dayonedesign.org
Earthen Built Kata Palano	BC, Canada	earthenbuilt.com
Endeavour Centre Chris Magwood, Jen Feigin	Peterborough, ON, Canada	endeavourcentre.org
Firespeaking Eva and Max Edleson	Deadwood, OR	firespeaking.com
Flying Hammer Productions Lydia Doleman	Southern Oregon	theflyinghammer.com
Frank Meyer	Austin, TX	thangmaker.com/natural.htm
From These Hands, LLC Sukita Crimmel	Portland, OR	sukita.com
House Alive Coenraad Rogmans, James Thomson	Jacksonville, OR	housealive.org
JRA Green Building	Portland, OR	jragbc.com
New Frameworks Natural Building	Burlington, VT	newframeworks.com
Peggy Frith	Winlaw, BC, Canada	peggydinah@yahoo.com
Proyecto San Isidro Alejandra Caballero	Tlaxco, Mexico	proyectosanisidro.com
Tactile Plastering Tracy Thieriot	Mendocino County, CA	tactileplastering.com/about.php
The Natural Builders	Bay Area, CA	naturalbuilding.com
Vertical Clay	Bay Area, CA	verticalclay.com
Vital Systems Tim Owen-Kennedy	Mendocino County, CA	vitalsystems.net

Appendix D:
Building a Compacted Gravel
(On-Grade) Subfloor

THE COMPACTED GRAVEL SUBFLOOR is a good substitute for a concrete slab when building a new structure from scratch. Concrete is an energy-intensive product, so minimizing its use is an important goal of environmentally conscious building. Compacted gravel subfloors have different structural properties than conventional concrete slabs and may not be appropriate for all situations. Consult with a structural engineer to ensure the subfloor and foundation are designed properly.

A compacted gravel subfloor.

(CREDIT: JAMES THOMSON)

Materials

Drainage gravel: Usually round, 1 to 1½" river rock, available from a sand and gravel supplier.

Vapor barrier: 6-millimeter plastic polyethylene sheeting, usually clear or black.

Insulation: Specific R-values and materials may be required if building code regulations are being followed; consult with the local building department or concrete contractor. Usually a high-compressive-strength rigid foam insulation is used, the same type designed to go under concrete slabs. For other insulation options, see Chapter 6.

Gravel Mix: The top layer is made of a hard-packed gravel and clay mix. Often this is referred to as roadbase. Another option is to use ¾" minus mixed with some site soil or imported clay.

Waterproof tape (duct tape)

Water

Tools: shovels, rake, hoe, hand tamper, plate compactor (optional, but helpful!), long level, 8' 2x4, laser level

Building the subfloor

Excavate the earth inside of the foundation footprint down to 12–18 inches below the desired finish floor level, or build the foundation stem wall up high enough to account for the thickness of this technique. Mark the desired subfloor level all the way around the inside of the foundation with chalk or a wax pen. Use a spirit level or laser level to be sure the line is level.

Compact the subsoil with hand tampers or a plate compactor. In especially wet environments, the subsoil surface can be graded to drain water toward a particular spot and out to daylight under the foundation. Lay drainage tubes (3"–4" perforated flexible pipe) on top of the compacted soil if this is a concern.

There are four layers that will go in your subfloor: the drainage layer (gravel), a vapor barrier, the insulation layer and the compacted gravel (roadbase). Plan in advance how thick each layer should be to bring your subfloor to the desired height.

Drainage layer: Over the compacted subsoil (and drain pipe, if used), lay 4"–8" of drainage gravel. Spread it evenly with a rake. Compact with the plate compactor or hand tampers.

Vapor barrier: Spread vapor barrier over the gravel. Run the barrier up three or four inches at the edge and tape to the foundation with waterproof tape. Tape any seams.

Insulation layer: Lay the insulation down to cover the entire floor all the way to the edges and tape the seams. The insulation should be flat and stable on the gravel, without voids that can create soft spots. If foam sheets are being used, a good trick is to place a layer of sand under the insulation and on top of the vapor barrier, to make it easier to create a level surface for the sheets.

Compacted gravel layer: Dump and spread the roadbase or ¾" minus over the insulation. If over insulation, use at least three inches of gravel, up to a maximum of six inches; if there is no insulation, use two to four inches. If the material is dry, spray with water as it is being added so that it is consistently moist. Use a laser level or spirit

Spreading and leveling gravel.

(Credit: James Thomson)

level and a 2x4 as a straightedge to make the subfloor flat and level. Add enough material to bring the floor up to (or slightly above) the desired subfloor level marks made earlier.

Compact the roadbase thoroughly. A plate compactor, if available, will work fairly quickly, but hand tampers will also work with enough pounding. Turn the compacting into a fun community experience by having a stomp dance party with friends and neighbors! Continue compacting until the surface is solid and flat.

Using a plate compacter to compact a gravel subfloor. (CREDIT: RON HAYS)

Using the power of many feet (and music) to compact a gravel subfloor.

(CREDIT: COENRAAD ROGMANS)

Check for level again. Rake down high spots and fill in any low spots; tamp again.

Allow the subfloor to dry thoroughly. This could take several weeks; thus it is best to build the subfloor early in the construction process so it can dry while the rest of the building is happening. Once it is dry, it is ready for a finished earthen floor.

The surface may require some touch-up prior to pouring the finish layer, to fix any divots or high spots created while the other work was being done.

Checking for level on a gravel subfloor. (CREDIT: JAMES THOMSON)

Appendix E:

Oil and Wax Safety

U SING LINSEED OIL AND THE SOLVENTS mixed in it are the most hazardous part of installing an earthen floor. The oils and solvents are harmful if ingested in large quantities. If swallowed, do not induce vomiting. Call a poison control center or a physician immediately. Drink small amounts of milk. If oil gets into the eyes, flush with water for ten to fifteen minutes. Consult a physician if irritation persists. Prolonged skin contact may cause drying. Consult a physician if irritation persists. Wear safety glasses and oil-impermeable rubber gloves during application.

Avoiding linseed oil fires

Linseed oil is also flammable, and under the right conditions it could spontaneously combust. When linseed oil dries, it releases heat as a by-product of the oxidation reaction. This reaction speeds up as the temperature increases, producing even more heat. This is not a problem when linseed oil is drying on an earthen floor (or on woodwork), as the air can easily absorb the released heat. The combustion risk comes from oil-soaked rags that have been crumpled up and disposed of before they are dry. The oxidation reaction can happen very quickly on the rags because they have a lot of surface area (and are thus exposed to a lot of air); if the rags are crumpled up, the heat cannot be released into the air and can build up, eventually reaching the oil's flash point. This avoidable situation has been responsible for many fires, including a major 1991 blaze in a highrise in Philidelphia that killed three firefighters.[12]

The solution to this problem is to let oily rags dry thoroughly before disposing them. Dry them flat outside, by hanging them on a line or laying them flat on the ground and weighting the corners with stones so they do not blow away. Oily rags can stain the surfaces they are laid on, so choose a drying location where this is not a concern or dry them on a piece of cardboard. If onsite drying is not possible, wet the rags with water and store them in a metal container to transport them to a suitable place to dry. Rollers used for oiling should be dried in a similar manner. Once dry, the rags and rollers can be thrown away in household trash.

Working with solvents

The other safety concern with linseed oil is related to the solvents that are used to thin it.

An inappropriately disposed-of oil rag. (CREDIT: JAMES THOMSON)

Oil rags set out to dry. (CREDIT: JAMES THOMSON)

All solvents, even all-natural ones such as citrus oil, contain volatile organic compounds or VOCs. This is a term used to describe a wide range of compounds that have low boiling points and thus evaporate quickly. There are many naturally occurring VOCs, such as methane and isoprene (released by plants); others are man-made products. Health risks arise with prolonged exposure to accumulated VOCs, a result of high concentrations of the compounds and/or inadequate ventilation.

VOCs evaporate and disperse quickly, and are thus more of a risk during the application process. The greatest risk is to the installer, and to anyone who will be in the room before the solvents have had a chance to fully evaporate. When working with solvents, make sure there is adequate ventilation (open windows and use fans to blow air out). When using oil with solvents, wear a respirator with an organic vapor cartridge. Make sure the room is sealed off from other rooms where people may be working or living while the oil is drying. Under normal conditions, 48–72 hours should be enough time for the majority of the VOCs to evaporate (the oil will likely need longer to fully cure). Always consult with the manufacturer's safety recommendations for each product.

Wax: The recommended floor waxes (see Resources appendix) contain some linseed oil and solvent. While the risk of combustion and VOC release is smaller with the wax because the overall amounts of oil and solvent are small, the same safety precautions should be taken when using wax and disposing of wax-coated rags and rollers.

Notes

1 *Earth Architecture* by Ronald Rael
2 http://architecture2030.org/the_problem/
  problem_climate_change
3 *The Elements of Style: A Practical Encyclopedia of Interior Architectural Details*, by Stephen Calloway and Elizabeth Cromley
4 Adapted from *Inventory of Carbon and Energy*, University of Bath
5 Adapted from *Adobe and Rammed Earth Buildings*, Paul McHenry
6 Richter, D.D., Markewitz, D., *Bioscience* 1995; 45:600-609; and Paul, E.A; Clark, F.E., *Soil Microbiology and Chemistry*. New York: Academic, 1989
7 World Green Building Trends SmartMarket Report by McGraw Hill Construction (2013): construction.com/about-us/press/world-green-building-trends-smartmarket-report.asp
8 LEED: usgbc.org/leed
9 Living Building Challenge: living-future.org/lbc
10 Report Summary of a Freedonia industry study, Hard Surface Flooring reportsnreports.com/reports/117994-hard-surface-flooring.html

11 Paul Hawken, *The Ecology of Commerce*
12 Gordon J. Routley, Charles Jennings and Mark Chubb, Highrise
   Office Building Fire One Meridian Plaza Philadelphia,
   Pennsylvania

Index

If you have enjoyed *Earthen Floors*, you might also enjoy other

BOOKS TO BUILD A NEW SOCIETY

Our books provide positive solutions for people who want to
make a difference. We specialize in:

**Sustainable Living • Green Building • Peak Oil • Renewable Energy
Environment & Economy • Natural Building & Appropriate Technology
Progressive Leadership • Resistance and Community
Educational & Parenting Resources**

New Society Publishers

ENVIRONMENTAL BENEFITS STATEMENT

New Society Publishers has chosen to produce this book on recycled paper made with
100% post consumer waste, processed chlorine free, and old growth free.

For every 5,000 books printed, New Society saves the following resources:[1]

33	Trees
2,948	Pounds of Solid Waste
3,244	Gallons of Water
4,232	Kilowatt Hours of Electricity
5,360	Pounds of Greenhouse Gases
23	Pounds of HAPs, VOCs, and AOX Combined
8	Cubic Yards of Landfill Space

[1]Environmental benefits are calculated based on research done by the Environmental Defense Fund and
other members of the Paper Task Force who study the environmental impacts of the paper industry.

For a full list of NSP's titles, please call 1-800-567-6772 *or check out our website* at:
www.newsociety.com

new society
PUBLISHERS